International Conference on

OPPORTUNITIES AND ADVANCES IN INTERNATIONAL POWER GENERATION

18-20 MARCH 1996

Organised by

The Power Division and North Eastern Centre of the Institution of Electrical Engineers and the Institution of Mechanical Engineers in association with the

Institution of Nuclear Engineers
Northern Engineering Centre
Institute of Energy

Venue
Treveylan College, University of Durham

Author Disclaimer

Copyright and Copying

Conditions of Acceptance of Advertisements

Published by the Institution of Electrical Engineers, London.

ISBN 0 85296 655 5 ISSN 0537-9989.

Printed in Great Britain by Short Run Press Ltd, Exeter

Organising Committee

Eur Ing R E Sellix, Engineering Council (Chairman)
Eur Ing F J L Bindon, Consultant
Mr J R Cure, Consultant
Eur Ing R.J Dinning, Scottish Power
Dr W W C Hung, National Grid Company
Dr K W Ingle, Scottish Nuclear Ltd
Dr P S Lawless, Northern Electric
Mr P Stokes, Teesside Power
Mr M N Storey, Merz and McLellan
Mr C J Thody, James Scott Ltd
Eur Ing R W Tindale, Davies,Tindale & Associates
Mr P N Waller, Retired
Prof H W Whittington, University of Edinburgh
Eur Ing P J Willows, IMechE

Corresponding Members

Mrs M Ilar, ABB Network Partner AG, Switzerland
Mr W I Rowen, General Electric Company, USA

iii

CONTENTS

> The Institution of Electrical Engineers is not, as a body, responsible for the opinions expressed by individual authors or speakers

COMMERCIAL ENGINEERING

PRIMARY ENERGY

ENERGY AND CONVERSION

SYSTEMS AND APPLICATIONS

List of Authors

TRENDS AND FUTURE DEMAND FOR ELECTRIC POWER GENERATION

Dr. Hisham Khatib

Honorary Vice Chairman
World Energy Council, Jordan

INTRODUCTION

Electricity is versatile, clean to use, easy to distribute and supreme to control. As important, it is now established that electricity has better productivity in many applications than most other energy products[1]. All this led to the wider utilization of electricity and its replacement to other forms of energy in many uses. Demand for electricity is now growing globally at a rate higher than that of economic growth and at almost 1.5-2 times that of demand for primary energy sources. With the type of technologies and applications that already exist there is nothing to handicap electricity's advancement and it assuming a higher share of the energy market.

Saturation of electricity use is not yet in sight even in advanced economies where electricity production claims more than half the primary energy use. Other than for the transport sector, electricity can satisfy most human energy requirements. It is expected that, in few decades, almost 70% of energy needs in most industrialised countries would be satisfied by electricity[2].

Electricity has become a most important ingredient in human life, absolutely essential for modern living and business. Its interruption can incur major losses and create havoc in cities and urban centres; its disruption, even if transient, can cause tremendous inconvenience.

This paper will describe the world electricity scene, its new trends and orientation and the effect of all this on the global electric power generating system.

THE EVOLVING GLOBAL ELECTRICAL POWER SCENE

This conference is being held at a crucial stage in the global development of electrical energy. Major changes are taking place prompted by :
(1) Growing environmental concerns about the role of energy, particularly electricity production, in causing emissions which have local, sometimes regional and (may be) global effects with long-term detrimental implications. (2) The growing importance of natural gas (also LNG) as a source of energy, particularly for electricity production; this has enhanced the utilisation and development of the gas turbine and its derivative, the high efficiency combined cycle plant, at the expense of the capital intensive traditional steam power stations. (3) Restructuring of the electricity supply industry and the growing importance of competition; this only reinforced the growing trend for more efficiency and less risk through investing in smaller sets with short lead times. (4) Stringent regulations (environmental and otherwise) and wayleaving conditions which increased the cost of new facilities and led to frenzy attempts to rehabilitate, refurbish and repower existing facilities.

Most of these considerations are going to have beneficial effects on the electricity generation sector. They can lead to cleaner and leaner production facilities, more rational and efficient use as well as fostering competition that assists in more efficient use of resources, thus bringing about more benefits to the consumer.

Opportunities and Advances in International Power Generation, 18–20th March 1996,
Conference Publication No. 419, © IEE, 1996

All this, however, is going to be slow. Energy systems, particularly electricity systems, have huge inertia; they are highly capital intensive and live for a long time. This, of course, delays reaping the beneficial results of some of these developments. Such delays have also been assisted by slow technological change in the way electricity is being produced, transmitted or distributed. Looking ahead two or three decades into the future, no technological breakthrough, which can revolutionise the way that electricity is produced or conveyed, is anywhere in sight. New and renewable energy sources (other than hydro) have not lived-up to their earlier promise or our expectation. Their economic development is, however, essential to provide a viable electricity source to the hundreds of millions who still have no accessibility to a commercial energy source or electricity supply. Such sources will of course take a dramatic boost in the future if the present worries about global warming are reinforced by more substantive arguments or proofs. It is not possible, in such a relatively short paper, to cover all the relevant technical and welfare issues surrounding electrical power, this was undertaken in more comprehensive work such as that of reference 3.

MAJOR TRENDS IN THE ELECTRICITY SCENE

The electricity supply industry, because of its inertia and institutional set up, has resisted change for decades . During the last few years, however, tremendous developments are gaining momentum . New trends in the global electricity generation have already been mentioned in the introduction, the most important among these are detailed below

1) Emphasis on efficiency: Prompted by environmental considerations, attempts to restrict emissions of local and regional detrimental emissions (SO_2, NO_x and particulate matter) and possibly global weather forcing green house gases (CO_2) , the ESI placed a lot of emphasis on efficiency. The generation efficiency, of less than 40% for new plant in the eighties,

has now risen to over 50% (possibly 60%) in case of combined cycle generating plant and over 42% (possibly 46%)[4] for thermal generating plant. Efficiency as means of limiting emissions has become the main issue of emphasis of the industry.Although SO_2 and NO_x can be controlled by emission control facilities, however, CO_2 can not. Improved efficiency contributes to reducing CO_2 emission. This emphasis on efficiency naturally led to wider adoption of combined cycle technologies.

2) Growing importance of Natural Gas and Gas Turbines: Natural gas contributed very little towards electricity generation in the past . Recently, with the growing emphasis on environmental cleanliness and rapidly growing global reserves, natural gas has become the favorite fuel for the ESI - it is clean, relatively cheap and abundant.

The growing utilisation of natural gas was accompanied by wider adoption of gas turbines and their derivative the combined cycle gas turbines (CCGT) as the favorite facilities for electricity generation. It is expected that in the next few years, larger proportions of all orders for new generating plant will be gas turbines based[5].

Natural gas suffers from limited availability in many countries, therefore cross boundary natural gas pipe lines are growing in numbers with long term delivery contracts becoming increasingly popular. To counter this trend, of generating utilities utilising gas as the clean fuel for electricity generation, prompted the development of coal gasification (IGCC), since coal is the most abundant fossil fuel and the favourite fuel for electricity generation for decades.

3) Shortages of capital: Because of its rapid development, the ESI is claiming an increasing share of gross capital formation (capital investment in infrastructure and services), particularly in developing countries.

This led to many developments the most important of which is : i) the introduction to the market of independent power projects (IPPs),

ii) low capital cost of facilities (gas turbine based) to limit investment, and iii) limited short lead times (also gas turbine based) to minimize uncertainties and risks.

4) Independent Power Projects (IPPs): Independent Power Projects (IPPs) are defined as "typically limited-liability, investor-owned enterprises that generate electricity either for bulk sale to an electric utility or for retail sale to industrial or other customers". IPPs are increasingly becoming an important source of power production in many industrialised countries(particularly the United States where their contributions to new capacity exceed that of regular utilities) . Few DCs faced with both power and capital shortages are now turning to IPPs to fill the gap. Ownership can take many forms the most important of which is BOO (build, own, operate), BOOT (build, own, operate, transfer).

The IPPs can coexist with the state owned utilities in many countries, but they need to be assured of few things : a power purchase agreement, secure fuel supplies and ability to convert currency and repatriate profits . IPPs which are investor operated can have an advantage of higher efficiency and better operation and management and thus higher availability . This can contribute to the security of supply and provide an example (and competition) with state utilities to improve their performance .

Finance to IPPs can be both foreign and local depending on the country and the development of its financial market. Foreign finance in case of developing countries, is usually through loans but can also involve export credit agencies and bilateral and multilateral lending agencies. Local financing can be from loans or participation by local banks for issue of bonds and other financial instruments .

The presence of an IPP project helps towards development of the local financial market. The presence of IPPs also greatly contributes towards restructuring of the power sector and the development of a regulatory body.

However IPPs investment, in most DCs, need guarantees .

A major problem in case of IPPs is the cost of financing. Because of the risks involved the project sponsors expect a net return usually 18-20% on their investment and after meeting all operational and other costs and loan repayment. This is a very high cost of finance which some countries, particularly most developing countries (DCs), cannot afford. Low income DCs in particular are used to the International Development Agency (IDA) and similar loans with long term repayments (20-30 years with 5-12 years grace period) at very low annual rates of interest (0.75-1%) . They cannot possibly afford the high financing costs associated with these new schemes, also they do not have the sophisticated local financial markets and institutions that can assist in co-financing and mitigating some of the risk . The World Bank recently developed financial instruments and institutions that can assist in reasonable financial risk mitigation to investors. These include : the Multilateral Investment Guarantee Agency (MIGA), Enhanced Cofinancing Operations (ECO) and also partial investment by the International Finance Corporation (IFC)[6] .

5) Demand Side Management: Demand Side Management (DSM) in an electrical power system have many aims. The most important is limiting the growth of peak by changing the pattern of electricity use ; thus differing the need for capacity expantion. Utilities that suffer from capital shortage to expand the system and from capacity shortages, find in DSM a very convenient way of integrated planning that will allow them to more efficiently utilise their limited resources. DSM also has other aims of enhancing conservation and improving efficiency in use. Conservation aims at rationalizing electricity consumption and limiting its over use.

Many generating utilities are finding it wiser to embark on DSM rather than on investment in system expansion[7].

Imaginative and efficient electricity pricing is the best tool for discouraging the use of electricity during peak hours. Consumers should be discouraged to use electricity during peak hours by the high tariff that reflects the long run marginal cost (LRMC) for electricity consumption during peak demand hours. They also need to be encouraged to conserve and rationalise their use and to shift their consumption to off-peak hours through an attractively low tariff that reflects the relative low system cost, as well as location and class of consumer and his type of load.

Not all DSM techniques rely on pricing. Working with consumers and equipment manufacturers can contribute to achieve the aims of rationalising the peak demand and improve the load factor. Consumer information and utility consumer links and dialogue are weak in many countries. They have to be developed in order to achieve the DSM goals. One of the methods of achieving this is to develop the Electricity Services Consumer Centres.

6) Restructuring of the Electricity Supply Industry:

Due to capital shortages and need for government finance and guarantee to loans and investment in the sector, governments dominated the electricity sector in most countries for many decades. Governments are not the best institution to run utilities. Government control, although it assisted in procuring finance to the electricity sector in the past, however it did not greatly contribute to productivity, efficiency and particularly to the future financial health of the sector particularly self-financing capabilities.

All this calls for change; for restructuring to reduce government control, competition in the market that leads to efficiency, and also outside finance (private as well as foreign). It is not possible to shed off government control which has been there for decades, this may also cause a lot of political resistance. Therefore the transition from full government to less government control has to be gradual. There is a need to create regulations that encourage the introduction of independent power projects to the sector, thus attracting private capital. Also to establish autonomous companies and corporations to run the sector.

FUTURE DEMAND FOR ELECTRICITY GENERATION

Electricity will continue to grow at figures higher than energy growth and economic growth. Most of the growth will be in DCs where electricity will be introduced to large sections of the population and will be utilized to fuel industrial and economic growth. In industrialised countries electricity will continue to substitute other forms of energy, thus its production will continue to rise in the future at rates higher than that of primary energy but at lower rates than those of the past; due to DSM, efficiency in utilisation and saturation of certain markets.

Estimations were made to predict future global electricity. These were detailed in references 8 and summarised in Table 1 below. They are based on the assumption that industrialised countries will maintain their economic growth at about 2.5-3% annually, while DCs will continue to have a higher growth, no less than 4.5% annually during this decade and dropping slightly thereafter.

From these predictions it is clear that the global electricity demand will grow at an average annual rate of 3.0-4.3% over the 1990s and will reach 16 000 TWh by the year 2000. Growth at a slightly lower rate will take place in the early part of the next century, with global demand estimated at around 22000 TWh and 28000 TWh in the years 2010 and 2030, respectively; with the rate of growth in DCs at 2-3 times that of the rest of the world (OECD, Central and East Europe). By the year 2030, the DCs which presently only produce 23% of the world's electricity, will be claiming half of the production. Electricity fuel needs, which are presently only 36% of total primary energy demand, will increase to more than 40% by the year 2000. By 2030, then, electricity will be claiming more than half of the world's total primary fuels.

TABLE 1- <u>Future Electricity Generation (TWH)</u>

	1990	2000	2030
OECD & others	9 020	11 150	14 200
DCs	2 680	4 500	14 200
World	11 700	15 650	28 400

FUTURE ELECTRICITY GENERATION FUELS[8]

The major fuels which now dominate the generation of electricity (oil, natural gas, coal, hydro and nuclear) will continue yet for many years to come. Owing to its high price, crude oil and its products will continue to have its share in electricity generation diminished in proportion and, most likely, in absolute terms. Its main utilisation will continue to be in OPEC countries, and also in small DCs which have no local energy sources. Natural gas is an ideal fuel for electricity generation and, correspondingly, it would increase its present share(15%) of the electricity market. The prospects of its assuming a much larger share will be enhanced in case of future energy/carbon taxation and environmental legislation. Coal is the world's major fuel for electricity generation and is going to remain so for many years to come. Its technologies are well established, prices are cheap, and reserves abundant. Regardless of its pollution problems, it will benefit from the fact that most of the growth in electricity generation will be in DCs which are relatively rich in coal, like China and India. These two countries where more than one third of the world population live will continue to depend mainly on local coal for electricity generation.

The future of nuclear energy, regardless of its established technologies and absence of gases emission, is handicapped by limited public acceptability, large size of units, and huge investment requirements. It must also be realised that its energy utilisation is limited to electricity production and that the future growth in electricity requirements will mostly be in DCs which, often , cannot meet the investment and technology requirements of nuclear energy. Therefore, for the foreseeable future, and other than for unpredictable environmental events (surprises), nuclear energy contributions towards global electricity requirements will, most likely, not increase. At best, it will be maintained at its present contribution of around 17%

As in the case of nuclear energy, hydroenergy can only be utilised for the production of electricity. However, in contrast to nuclear energy, hydroenergy has limited resources with most of the large economical sites already utilised. Therefore, its future proportional share in electricity production would decrease. During the 1980s, hydroenergy, grew at a handsome 2.3% annually, but this has limited future prospects towards sites which demand higher capital investment and which have greater environmental impact.

The contribution of new and renewable energies in the form of solar, photovoltaic, wind, geothermal etc., are very limited and will remain so. Even if there is a technological breakthrough in the future (which is not evident yet), its effect will only be very gradual and limited, given the enormous stock of existing generation facilities.

References:

1. Gellings C, 1992, "Saving Energy with Electricity" , EPRI Report, Palo Alto, U.S.A.

2. Gerholm T, 1991, "Electricity in Sweden - Prospects to the years 2050", Vattenfall, Sweden.

3. Khatib H., Munasinghe M, 1992, "Electricity, the Environment and Sustainable Development", World Energy Council 15th Congress, Madrid, Spain

4. EPRI, 1955, <u>Power Engineering International</u>, 5,21-26

5. EPRI,1955, <u>Gas Turbine Journal,</u> 3,20

6.   The World Bank, 1955, <u>Energy
      Themes</u>, FPD Energy Note No 2,
      Washington DC, USA

7.   The World Bank, 1993, "Energy
      Efficiency at Conservation in
      the Developing World",
      Washington D.C., USA

8.   Khatib H, 1993, "Electricity in
      the Global Energy Scene", IEE
      Proceedings-A, 1, 24-29.

COMMERCIAL ENGINEERING: FINANCE

ANDREW KINLOCH, VICE PRESIDENT, UNION BANK OF SWITZERLAND

Introduction

Commercial Engineering: Finance is a huge brief. Funds for an engineering project, be it a road, railway, bridge, oil or gas rig, paper mill, mine, telecommunications network or power station, can come from any of the following sources:

- The project's sponsors inject the required funds, ie use their own balance sheet;
- New equity is raised from the public;
- Leasing, which enables tax deductions to be utilised sooner;
- Capital markets instruments which are gradually winning acceptance, whereby relatively inflexible bonds are issued to investors;
- Export credit agencies such as the UK Government's ECGD / multilaterals such as the World Bank provide fixed, long term funding and take political, and sometimes commercial risk;
- A number of banks share in the project risks by providing bank debt to the project with only limited recourse to the sponsors of that project.

You will therefore no doubt be relieved to hear that I intend to focus on only one type of project and the last of the types of finance, namely a power plant funded primarily by limited recourse bank debt, i.e. the Independent Power Project or IPP. I would like to do so because it is in this field that new opportunities are arising and where engineers and financiers are being required to work together to address new issues, new risks and, we trust, new rewards.

Typical Independent Power Project ("IPP") financing structure

First, we need to understand what an IPP looks like. This slide [attached] shows the contractual relationships between the various parties needed to make an IPP happen. In the centre is a special purpose project development company, funded by some sponsors' equity, mostly our bank debt and perhaps some of the other sources referred to just now. A contractor builds the plant over two to three years, typically under a turnkey contract. Upon completion of construction, fuel supply, operational capability and sales commence pursuant to long term contracts, say 15 years. Frequently, some or all of the sponsors act as one of these counterparties too, ie they sponsored the project so as to create themselves a new customer.

The special purpose company at the centre of the contracts is our borrower, which draws down the funds required for construction over two to three years then repays our debt over the operational term of 15 years. It does so out of cash generated from only the project so we are relying on the project to be first built then operated successfully if we are to get our money back as intended. Once the banks have been paid off, the sponsors have a debt-free asset, still worth a considerable sum. In practice, however, once the principal risks of construction and the first year or two of operations have been safely overcome, we would expect to be refinanced so as to achieve lower pricing or less onerous covenants imposed on the borrower.

Issues

Three of the most topical risk considerations are:

Political risk

IPP opportunities are often spawned upon privatisation of nationalised industries. Privatisation requires the people doing the privatising, ie various departments in Government, to understand what risks can be transferred into the private sector; and what responsibilities, they will need to

Opportunities and Advances in International Power Generation, 18–20th March 1996,
Conference Publication No. 419, © IEE, 1996

retain. eg a regulatory role. Consistency of policy in respect of mergers and monopolies, tax regime and that regulatory framework is necessary in an industry which has to take a long term view of the investment decision. By the same token, it requires those of us in the private sector to recognise that many of these concepts are novel and politically sensitive.

The industry is often privatised gradually so both sides need particularly the purchaser of the power to be creditworthy both whilst still Government-owned but also in the long term once privatised. This transition has been successfully managed in Northern Ireland then Portugal. On the other hand, Italy is currently deciding how it wants to privatise its power industry. And India is changing its mind, as Enron, without a local partner to ease the liaison, is finding to its cost at Dabhol.

Completion risk

Turnkey construction contracts are typically structured so that, at the risk of mixing metaphors, the first slice of any pain incurred by a project is borne by whichever party was responsible for the pain. Contractors can consequently find themselves paying out substantial liquidated damages. However, if the liquidated damages are used up and still the pain continues, as it were, then the project feels it and once an equity buffer is used up, so do the banks. Now, several IPPs in the UK have been, or are going to be, completed late. In addition to the actual project contracts, the banks are consequently reassessing contractors' ability to deliver on their obligations set out in those contracts. This includes their project management skills, their approach to the development of new technology and their overall commitment to the project.

One interesting consideration is how does the development company / borrower behave when one of its sponsors is also the contractor?

Technology risk

We are seeing the implementation of several new technologies, such as renewable energy sources across Europe and coal and tar gasification in Italy. This is often driven by political motives. If the technology doesn't work, though, we banks are going to find ourselves with a borrower unable to repay us. We therefore take extensive advice from engineering consultants that the technology contemplated will work as designed on the intended scale and with the intended efficiencies, output and reliability. We next structure the financing such that if it does not perform to plan, the first "slice of pain" is again taken by the party responsible (either the contractor or the operator), the next slice by equity and then the rest by us.

Appetite for financing

As is so often the case, we spend most of our time in negotiations worrying about all the potential bad news. Let this not obscure the good news, however. On the one hand, there is huge demand for power plants worldwide (especially in developing countries) but with a local public sector which often cannot afford such capital expenditure. On the other hand, major developers have the construction and operating expertise to export but are wary of new risks in new countries. Financiers prepared to share some of the risks with the other contracting parties are enabling projects to be built which would otherwise not be built and this is good news for both engineers and for financiers.

Questions

9

Typical Power Generating Project - Counterparties and Contracts

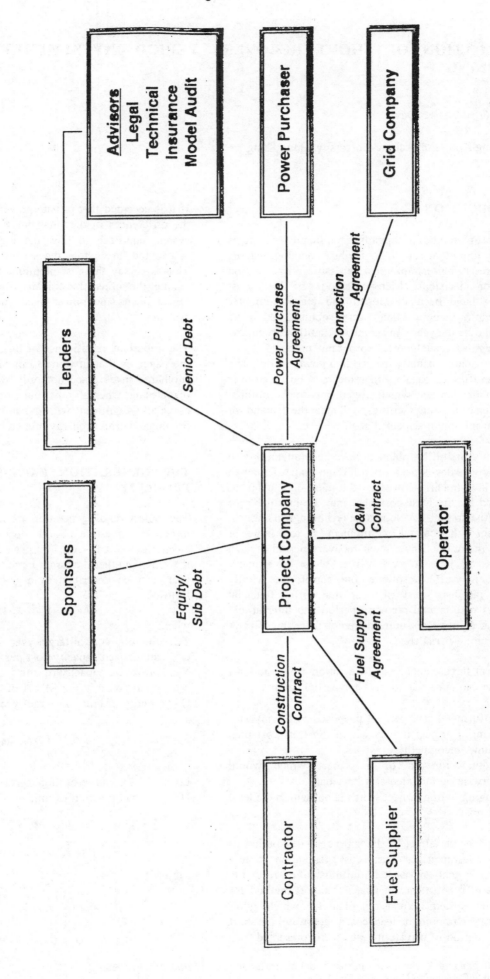

EVALUATION OF SHORT RECOVERY PERIOD INVESTMENTS ON UTILITY PLANNING

Y. Q. He and A. K. David

The Hong Kong Polytechnic University, Hong Kong

INTRODUCTION

Least cost planning for expanding public power systems plays an important role in achieving economic benefits while preserving technological necessities in generation expansion decisions. Many methods of least cost planning have been developed and many computer packages[1,2], some of them very sophisticated, have emerged over the years. Investments traditionally follow an integrated development program prepared by a planning agency, usually located in a power utility. The plan identifies the capacity increments to be added over the planning horizon which, together with the optimal utilization of existing facilities, will meet the demand for power at minimum expected cost.

In recent years utility planners have been attempting to attract private developers, often foreign, to participate in power investment. However, the next-best plant as identified by the least cost plan may be unsuitable or unattractive to such developers for technological reasons, or because the capital requirement is too large or because the normal investment recovery period of such plant is too long to suit private sector investment. Investors may offer unsolicited bids for projects wholly outside the least cost plan or may offer financial packages with capital recovery period and interest and return on equity rates quite different from the utilities own planning parameters.

In general three categories of deviation from the least cost plan can arise due to investor participation

(a) The financial arrangement proposed by the investor making a project that is within the least cost plan become uncompetitive
(b) Bringing forward of a project which would otherwise be included at a later date
(c) Bringing in of a project which is not within the least cost plan

In this paper the effect of the short recovery period of private investment on least cost generation system extension is analyzed as it is the most often recurring example of type (a) problems. A trade-off method for generation system expansion planning which gives consideration to both the least cost strategy and the short recovery period of private investment is presented.

If it is accepted that private power is necessary despite the differences in the financial arrangement, a whole-system approach to how the higher costs are to be distributed throughout the system must be developed. That is to say the cost premium has to be shared out among the other traditionally financed public sector power plants and power plant already existing in the system.

The important significance of this, as will be shown in the paper, is that in tariff formulation recovery of the surcharge takes place not only when the private unit operates but when all other units operate as well. This is especially significant with dynamic tariffs, time-of-day pricing and pricing energy interchange between systems.

THE GENERATION EXPANSION PLANNING PROBLEM

The objective of generation expansion panning is to minimize investment and operation cost to meet electricity demand over a target period. This objective, if we assume that the target period consists of only one year, can be described in a mathematical model as follows,

$$Min \ TC = \sum_i (C_i \cdot X_i + b_i \cdot Q_i) \qquad (1)$$

where
TC: the total cost in target year
C_i: annualized capacity cost per MW of unit i
X_i: operation capacity of unit i
b_i: operation cost per MWH of unit i
Q_i: energy of unit i in target year

$$Q_i = \int_{D_i}^{D_{i+1}} L(u) \ d(u) \qquad (2)$$

L(u): LDC curve of target year
D_i: load position of unit i

$$D_1 = 0, \qquad D_i = \sum_{j=1}^{i-1} X_j, \qquad (3)$$

$$i = 1, 2, \cdots, n+1$$

subject to,

$$\sum_{i=1}^{n} X_i \geq L^m \qquad (4)$$

$$X_i \geq 0 \quad (i=1, 2, \cdots, n) \qquad (5)$$

$$X_i \leq Y_i \qquad i \in I_1 \qquad (6)$$

$$Q_i \leq \overline{Q_i}, \qquad i \in I_2 \qquad (7)$$

Opportunities and Advances in International Power Generation, 18–20th March 1996,
Conference Publication No. 419, © IEE, 1996

where

L^m: system peak load in target year

Y_i: capacity of unit i (existing units only)

$\overline{Q}_i$: annual energy limit of unit i (eg. hydro)

I_1: set of units with capacity constraint (eg. existing units)

I_2: set of units with energy constraint (eg. hydro)

Although only one target year is used here for simplicity, the method has to be extended to multi-year planning. The full computer algorithm that has been developed covers a multi-year planning strategy, but for clarity the discussion below is confined to a single target year.

Since the short recovery period problem is to be discussed in this paper, a representation of the recovery period in the capacity cost C_i has to be modelled. The following relationship will be employed to represent the equivalent unit capacity cost

$$C_i = F_i \cdot \frac{\alpha}{1-(1+\alpha)^{-T}} \cdot \frac{1}{X_i} \qquad (8)$$

where

F_i: total investment in respect of unit i

α: yearly interest rate

T: investment recovery period

Typically, the investment recovery period (T) is 25 years, and the interest rate (α) has been between 8 - 16% in the past but will be higher with investor owned plant both due to higher returns on equity and the shorter capital payback period.

Using the Kuhn-Tucker theorem a series of K-T conditions can be derived from the model[3,4] in eqn(1)-(7). For a new thermal unit,

$$C_i + R_i - \mu - \mu_i = 0 \qquad (9)$$

where

μ: K-T multiplier of constraint (4)

μ_i: K-T multiplier of constraint (5)

$$R_i = \sum_{l=i}^{n-1} (b_l - b_{l+1}) \cdot L(D_{l+1}) \qquad (10)$$

For an existing thermal unit i,

$$\lambda_i + R_i - \mu - \mu_i = 0 \qquad (11)$$

where

λ_i: K-T multiplier of constraint (6)

Comparing (9) with (11), we see that the role of λ_i is similar to C_i. The point can be illustrated as follows. Suppose there is a new (candidate) thermal unit with capacity cost C_i and operation cost b_i, say after optimisation the required optimal capacity is X_i and required optimal energy is Q_i. Now, suppose instead that there is an existing thermal unit with capacity $Y_i = X_i$ and operation cost b_i. After optimisation it will be found that

its optimal energy must be Q_i and λ_i must equal C_i. Hence λ_i reveals the capacity value of the existing unit and is called the shadow capacity cost.

The K-T multipliers μ, μ_i for all i, are zero if the corresponding constraints are not binding and positive if binding. They too indicate the shadow cost or value of the corresponding constraint.

From the K-T conditions it can also be shown that for a new hydro unit h, the optimal loading point on LDC time-axis is given by,

$$\tau_h = L(D_h) = \frac{C_{h-1} - C_h}{(b_h + \gamma) - b_{h-1}} \qquad (12)$$

or,

$$\tau_{h+1} = L(D_{h+1}) = \frac{C_h - C_{h+1}}{b_{h+1} - (b_h + \gamma)} \qquad (13)$$

where γ is the K-T multiplier of constrain (7). For a system which considers only new thermal units, the optimal loading point on the LDC time-axis is,

$$\tau_{i+1} = L(D_{i+1}) = \frac{C_i - C_{i+1}}{b_{i+1} - b_i} \qquad (14)$$

Comparing (13) with (14), we see that $b_h + \gamma$ is equivalent to b_i. Since $b_h \approx 0$, γ is equivalent to b_i. That is, suppose there is a new thermal unit with capacity cost C_i and operation cost b_i, and say after optimization the required optimal capacity is X_i and optimal energy is Q_i. Now suppose instead that a new hydro unit with capacity cost $C_h = C_i$ and limited energy Q_i is considered. After optimization, it will be found that its optimal capacity X_h will be equal to X_i and that γ will equal b_i. Hence γ reveals the energy value of the new hydro unit, called the shadow energy cost.

The Value Added Generation Expansion Planning method of Ref.(1) was used to obtain the optimal solutions in this paper. But the choice of algorithm, provided it can generate shadow prices, does not fundamentally affect the discussion in this paper.

EFFECT OF SHORT RECOVERY PERIOD

A case study showing the effect of short investment recovery period on least-cost generation expansion planning is discussed below. The load data and existing thermal unit data are those of the 1990 Shanghai power system, but the new thermal unit data are typical assumed values. The data is shown in Tab-I, Tab-II and Tab-III respectively. (The exchange rate was about 5.4 Yuan=US$1.0 in 1990).

Table I Load Data

Peak load (MWH)	Energy (GWH)	Load Factor (%)
4102	30.14	83.9

Table II Existing thermal unit

Operation cost (Yuan/MWH)	Capacity (MW)	Avail. (%)
73.6	182	95
59.4	100	92
52.4	1200	79
51.2	740	87
36.8	255	93
36.8	377	89
32.6	277	89

Table III Data of proposed thermal units

No.	Operation cost (Yuan/MWH)	Cap. cost (Yuan/kW)	Avail. (%)	Max. Cap (MW*No.)	Attributes
1	90.0	200	87	200*1	peak
2	70.6	265	87	200*1	peak, private
3	50.1	312	79	600*3	middle
4	35.6	504	87	600*3	base, private

Table IV The economic indices of proposed private thermal units

No	Capacity (MW)	Total Investment (billion Yuan)	Recovery period (Years)	Interest (%)
2	200	4.66	25	10
4	600	18.55	10	10

Assume that unit 2 and 4 in Tab-III are earmarked as private investment projects whose economic indices are shown in Tab-IV. The recovery period of unit 4 is 10 years, and of all other units (1, 2, and 3) 25 years. The optimal expansion plan calculated by the value added method is shown in Tab-V. Two new units, unit 1 and 3 are accepted in the optimal extension plan. Unit 1 has an advantage in its lower capacity cost, and unit 3 in its lower operation cost with reasonable capacity cost. The two private investment unit, unit 2 and 4, are rejected from the optimal expansion plan. The reasons for the rejection of the two units are different. Unit 2 is rejected because it is inferior to units 1 and 3 in costs, on consideration of both capacity and operation. But the rejection of unit 4 is only because of its short recovery period (10 years), which raises its effective capacity costs as modelled by eqn.8. From the point of view of making best use of natural resources, it is desirable that unit 4 be included in the system expansion plan, but the short recovery period of the private unit is an financial constraint that excludes it from the plan. If its recovery period was the same as that of other new units (25 years), the optimal result is as shown in Tab-V. The unit is accepted in the optimal expansion and the other three units are all rejected. (The concept is shown in Fig-1 using unit k with capacity cost C_k or C'_k and operation cost b_k). The recovery period affects the competitiveness greatly and the relationship between the recovery period and cost and accepted capacity of this unit is shown in Fig-2. If the recovery period is less than about 12 years, the unit will be excluded from the least cost plan.

The problem of recovery period should therefore be separated from the least cost expansion problem. On the one hand, from the point of view of making best use of natural resources, least cost expansion planning should be based on a standard recovery period depending only on the technical attributes of the plant and regardless of financing arrangements. However, for some reason such as capital requirements, government encouragement or by market policy, a private project with a short recovery period may be enforced. If the private unit is accepted in the optimal expansion plan, the cost increase due to the shorter recovery period should be spread across the whole system, rather than charged to this unit only. The following trade-off method achieves this objective.

Table V Optimal expansion under 10-year/25-year recovery period of unit 4

No.	Optimal Capacity (MW)	Optimal Energy (GWH)
1	180 / 0	116 / 0
2	0 / 0	0 / 0
3	1609 / 0	11100 / 0
4	0 / 1643	0 / 12468

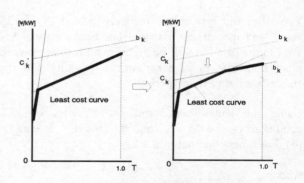

C_k & C'_k -- per unit Cap. cost under standard & short recovery period respectively

Figure 1 Illustration of the acceptance process of private unit k

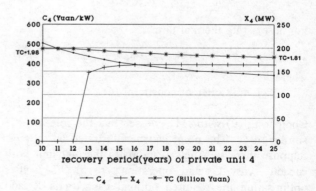

Figure 2 The relationships between recovery period and capacity cost (C_4), capacity (X_4) of the private unit 4 and total cost of the system (TC)

TRADE-OFF METHOD FOR LEAST COST STRATEGY WITH SHORT RECOVERY PERIOD

Mathematically, because of the participation of private units with short recovery period, the objective function needs to be expressed as,

$$Min\ TC = \sum_{i=1}^{n}(C_i \cdot X_i + b_i \cdot Q_i) + \sum_{j=1}^{m}\eta_j \cdot X_j \quad (15)$$

where

$$\eta_j = C_j^s - C_j \quad (16)$$

and,

C_j: annualized capacity cost of unit j under a standard recovery period

C_j^s: annualized capacity cost of unit j under a short recovery period (e.g. 10 years)

i: set of all units

j: set of short recovery period units

The expression consists of two parts; the first part,

$$TC^* = Min \sum_{i=1}^{n}(C_i \cdot X_i + b_i \cdot Q_i) \quad (17)$$

is the objective function under a standard recovery period. The optimal capacity of unit j is X_j^* and the total cost of the system is TC^* (the point "A" shown in Fig-3). The optimal result may be unattractive to private investors because they prefer a shorter recovery period. The second part $\Sigma \eta_j \cdot X_j$ is the surcharge caused by the shorter recovery period demanded by private investors. The traditional least cost strategy would set the optimal result of eqn.15 at point "B" shown in Fig-3. The total cost of the system would be TC^{**} and the optimal capacity of unit j would be $X_j^*=0$. Unit j is rejected due to its high capacity cost ($C_j + \eta_j$).

The key point in applying a least cost strategy is to maintain a level playing field so that all units can compete with each other. The surcharge is produced by financial and social constraints and should be spread across the whole system, rather than charged to some units only. Using this principle, we separate the surcharge from the objective function. First, optimization is made on the eqn.17, which is the first part of the objective function (eqn.15) where all new units are under a fair recovery period. The least cost TC^* and the optimal X_i^* (where i includes all j) are obtained. Then using X_j^*, the surcharge (SC) caused by the short recovery periods of some units (j) is calculated as follows,

$$SC = \sum_{j=1}^{m}\eta_j \cdot X_j^* \quad (18)$$

Finally, the total cost of the system is obtained as

$$TC = TC^* + SC \quad (19)$$

and corresponds the point "C" in Fig-3.

The surcharge SC must be charged to system capacity cost, while the optimal capacity, energy of every unit and merit loading order remain unchanged. This process makes the whole least cost curve float up by an amount ΔC, which is calculated as follows

$$\Delta C = \frac{SC}{\sum_{i=1}^{n}X_i^*} \quad (20)$$

ΔC should be added to capacity cost of all new units and shadow costs of all existing units and a set of modified capacity costs, under the short recovery period for new private units, should be derived for all units including existing units. That is, unit i modified capacity cost C" (or modified shadow capacity cost λ_i") is given by

$$C_i'' \ (or\ \lambda_i'') = C_i \ (or\ \lambda_i) + \Delta C \quad (21)$$

The process is shown in Fig-3 and C" (or λ_i") should be used for tariffication as described in the next section.

The results for the previous example have been calculated and listed in Tab-VI. For comparison, the results by least cost strategy under recovery period of 10 and 25 years are also listed. The total cost by the trade-off method is the highest among the three cases. The reason is the shift from point "A" on the least cost curve to point "C" in Fig-3. The apparent annualized capacity cost of unit 4 is 397 Yuan/kW rather than 504 Yuan/kW (the actual short recovery capacity cost) because the short recovery surcharge is shared by other units in the system.

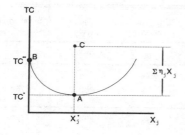

Figure 3 The effect of short recovery period on least cost plans

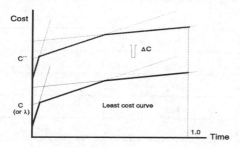

Figure 4 The sharing the surcharge among all units

EFFECT OF PRIVATE INVESTMENT ON MARGINAL PRICE

The least cost strategy is affected by private investment with a short recovery period, and inevitably, the marginal price is influenced. A marginal pricing method based on a least cost strategy was developed by Yen *et al*[5], where the system marginal capacity price (P_C) is

14

equal to the annualized capacity cost of the last unit in merit loading order if the last unit is a new one, or the shadow capacity cost of the last unit if it is an existing unit. The marginal energy price (P_E) is equal to the increase in system energy cost if one additional unit of energy is produced. A description of the marginal capacity price and marginal energy price computation in the presence of private sector power investment is provided in the Appendix.

Under the least cost strategy, the effect of short recovery period of private unit 4 on marginal capacity and energy costs is shown in Fig-4 for the previous example. The longer the recovery period of the unit, the lower the marginal costs.

With the proposed modification of the least cost strategy, the marginal capacity and energy prices (P_C and P_E) also change. The marginal capacity and energy prices P_C and P_E for the three situations discussed above are listed in Tab-VI. For a 10-year recovery period of private unit 4 (first line in Tab-VI), the last unit in merit order is new unit 1, with capacity cost C=200 Yuan/kW, so the system marginal capacity price P_C=C=200 Yuan/kW. For a 25-year unit with operation cost b=73.6 Yuan/MWH and shadow capacity cost recovery period of unit 4, the last unit in merit order is an existing (λ) of 152 Yuan/kW, hence P_C=λ=152 Yuan/kW (second line in Tab-VI). With the trade-off method, the additional capacity cost ΔC is 57 Yuan/kW. According to eqn.21, the modified shadow capacity cost λ"=209 Yuan/kW, so the marginal capacity price P_C=209 Yuan/kW. In this example this is the highest of the three situations, but this will not always be the case as can be appreciated from the above discussion.

The difference of marginal energy price (P_E) between 10-year and 25-year recovery period of private unit 4 is caused by the different composition of units. Because the compositions of units are the same in situations of 25-year recovery period and trade-off method, the system marginal energy price in these two situations are the same.

A PRACTICAL CASE STUDY

This example shows that the trade-off method applied to Shanghai generation system expansion planning for year 2000. There are 10 generation plants in the system with total capacity 5578 MW up to 1994. The data of the units in these plants are listed in Tab-VII. (The current exchange rate is 8.3 Yuan=US$1.0). The peak load and load factor in year 2000 are predicted to be 10360 MW and 66%. The power to be purchased from the middle and southeast China districts in year 2000 is 660 MW and 5780 GWH per annum. The candidate unit data is shown in Tab-VIII, where *ZB* (100*4 MW) and *SDK-2* (600*2 MW) are financed by non-government power development funds and require 8-year and 15-year recovery, respectively, instead of 15-year and 25-year, fair periods of these units. The optimal expansion plan is obtained by the trade-off method and listed in Tab-IX. The surcharge caused by the shorter recovery period amounts to 1.3 billion Yuan and is spread across the whole system making an added cost 114 Yuan/kW to capacity costs of all new units and shadow capacity costs of all existing units. The marginal capacity price and marginal energy price are 3373 Yuan/kW and 0.389 Yuan/kWH respectively.

Table VI Comparison of cost under three situations

Situation		Private project unit 4		Total Cost (Billion Yuan)	P_C* (Yuan/kW)	P_E (Cent**/kWH)
		Capacity cost* (Yuan/kW)	Capacity (MW)			
Least cost strategy	10 Years	504	0.0	1.98	200	7.25
	25 Years	340	1643	1.81	152	6.33
Trade-off method		397	1643	2.08	209	6.33

* Capacity cost and P_C are annualized rates. ** 1 Yuan = 100 Cents

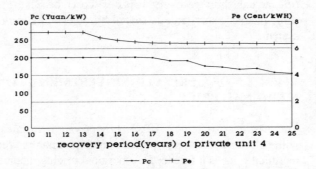

Figure 5 Recovery period of private unit 4 vs marginal capacity cost P_c and marginal energy cost P_e

Table VII Existing units of Shanghai system in 1994

Plant name	Unit type (MW*No)	Capacity (MW)	Operation cost (Yuan/MWH)	Avail. (%)
YSP	< 100	215.7	216	95.5
ZB	125*3	400	169	95.5
	< 100			
WJ	125*5	625	169	95.5
	< 100	325	216	95.5
MH	125*4,110*2	720	169	95.5
	< 100	9.8	216	95.5
NS	< 100	115	216	95.5
SDK-1	300*4	1200	154	91.0
SDK-2	600*2	1200	149	86.0
JS	< 100	276.5	216	95.5
GQ	< 100	228	216	95.5
JS-2	< 100	175	216	95.5

Table VIII Proposed units for year 2000 of Shanghai system

Plant name	Unit type (MW*No)	Cap. cost (Yuan/kW)	Operation cost (Yuan/MWH)	Avail. (%)
WGQ	300*4	4531	154	91.0
NS	50*1	3258	332	96.5
YSP	100*2	3258	332	96.5
ZB	100*4	3258/4463*	332	96.5
WJ	600*2	4574	149	86.0
SDK-2	600*2	4574/5260**	149	86.0
WGQ	900*1	4793	120	86.0
MHN	600*2	4574	149	86.0

* Under recovery period of 15/8 years. ** Under recovery period of 25/15 years

Table IX Optimal expansion plan using trade-off method

Plant name	Modified Cap. cost (Yuan/kW)	Optimal capacity (MW)	Optimal energy (GWH)
WGQ	4646	12000	6110
NS	3373	50	33
YSP	3373	200	116
ZB	3373	400	105
WJ	4689	1200	9040
SDK-2	4689	1180	8802
WGQ	4908	900	6780
MHN	/	0	0

CONCLUSION

A trade-off method is presented in this paper to give consideration to both the least cost strategy of the whole system and the short recovery period of private investment. The surcharge caused by the short recovery period of private units is separated from the least cost objective. The least cost strategy is applied on a standard recovery period for all new units, regardless of the financing arrangements, and the surcharge is passed on to the whole system rather than charged to some units only. The former gives consideration to making best use of natural resources, and the later is a trade-off method to allocate surcharge costs to all plants including non-short recovery plant, so as to achieve an economic balance.

The practical significance of this is that in tariffication the recovery of the surcharge takes place not only when the private unit operates but when all other units operate as well. This is especially significant with dynamic tariffs, time-of-day pricing and pricing energy interchange between systems.

REFERENCES

1. Yen M.S., Miao H.J., He Y.Q. and Li D.G, 1992, "Value Analysis of Generation System Power (VAGS)", Conf. Proc. SEPRI '92-1, New Orleans, Louisiana, 1-10

2. Jenkins R. T. and Joy R. S., 1974, "Wein Automatic System Planning Package (WASP)", ORNL-4954, Oak Ridge National Laboratory

3. Levin N. and Zahavi J., "Optimal Mix Algorithm with Existing Units", 1984, IEEE Trans. on PAS Vol. PAS-103, No.5, 954-962

4. Levin N. and Zahavi J., 1985, "Optimal Mix Algorithm with Limited-energy plants", IEEE Trans. on PAS Vol. PAS-104, No.5, 1131-1139

5. Yen M.S., He Y.Q. and Zhuang S.C., 1993, "Time-of-day Pricing Method Using Optimal Mix of Generation System", Conf. Proc. of CIRED, London, UK

Acknowledgement

A research grant from the Research Grants Council of Hong Kong is gratefully acknowledged.

Appendix

The well known the definition of marginal capacity price (P_C) is the increase in total capacity cost due to one additional unit of capacity in peak load. This additional unit of capacity demanded will be met by the last unit in merit loading order. If this unit is a new one, its capacity cost (C_n) is the increase in total capacity cost, that is $P_C=C_n$. If it is an existing unit, the increase in total capacity cost due to one additional unit of capacity committed in peak load equals the effective increase in capacity cost of this existing unit. The later is obtained as the Kuhn-Tucker dual variable of capacity constraint of the unit and is a normal output value available from the optimization program which has been used to obtain the results presented in this paper. It is named the shadow capacity cost (λ_n) of the unit, and we set $P_C=\lambda_n$. Hence the system marginal capacity price is,

$$P_C = \begin{cases} C_n, & \text{if } n \text{ is a new unit} \\ \lambda_n, & \text{if } n \text{ is an existing unit} \end{cases} \quad \text{(A1)}$$

where unit n is the last unit in merit order after optimization.

The system marginal energy cost can now be written as

$$P_E = \lim_{\Delta Q \to 0} \frac{\Delta TV - P_C \cdot \Delta X}{\Delta Q} \quad \text{(A2)}$$

where

ΔQ: increment of system total energy
ΔX: increment of peak load
ΔTV: increment of system total cost, including shadow cost, under optimal mix due to ΔQ

Eqn.(A2) is derived from the definition of marginal energy cost, $P_E=(\partial TV'/\partial Q)$, where $\partial TV'$ is the increment in total energy cost due to the additional energy ∂Q produced.

The associated capacity cost increment, $P_C \cdot \Delta X$, which is a first order approximation, should be excluded from ΔTV in obtaining P_E. The increment ΔQ and ΔX is set by marginally incrementing (modifying) the load duration curve and the calculations are repeated to obtain the ΔTV.

FUEL PURCHASE FOR MIXED FUEL GENERATORS

G. M. Bellhouse, H. W. Whittington

University of Edinburgh, Scotland

ABSTRACT

In the increasingly competitive UK electricity generation market, cost-effective fuel procurement will prove to be a critical factor in business success for UK power Generators.

This project aims to formalise the fuel purchasing decision process and develop decision reduction techniques. These will be incorporated into interactive decision support software specifically designed to maximise the effectiveness of fuel purchasing.

1. INTRODUCTION

Fuel procurement is the largest single item in the operating expenditure of any electricity Generator, and as such, is crucial to competition within the industry. In 1994 the UK electricity generating industry purchased 35.9 million tonnes oil equivalent of coal and 3.58 million tonnes of oil for the production of electricity [1].

Within most electricity generating utilities there is a constant need to have a balanced portfolio of fuel purchase contracts. Even within the purchase of any one particular fuel, several contract types are often maintained to ensure maximum flexibility of supply. The procurement of black fuels (coal and oil) is one such area and is exposed to a number of influences not experienced by other fuels. Commercial aims of fuel purchase include minimising system operating costs with regard to security, whilst respecting operational and legislative constraints.

2. BACKGROUND

Although it is fairly straight forward to purchase black fuels, the complexity and range of influencing factors make the effectiveness (or "goodness") of the decision itself much harder to assess. For the present study, initial work has concentrated on analysis of past data from one Generator, looking at any major variances from expectations and examining the influence of variation in other factors.

3. MERIT ORDER

In an increasingly competitive electricity market the struggle to become a base-load generator is set to become more intense. Being a base-load generator offers financial security in terms of guaranteed

generation and therefore shorter pay-back periods for capital investment of generation projects. Technical reasons for being base-load Generators include not having repeatedly to reduce output from stations, especially important for nuclear generation due to the risk of fuel poisoning.

With the projected increase in generation from gas, use of coal seems likely to decrease, moving from base-load generation to marginal generation. The consequences of this will be increasing significance of the magnitude of variation in coal use relative to its total use. As a result, planning of coal consumption and purchase will become increasingly critical to utilities, especially those who traditionally had a large dependency on coal. Figure 1 shows how the breakdown of electricity generation has changed in recent years.

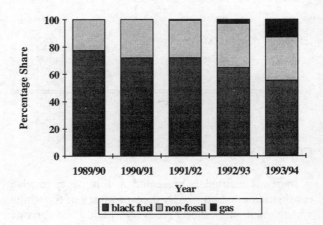

Figure 1: Breakdown of Electricity Generation, 1989-94 [2]

Figure 2 shows a typical load duration curve for an electricity network supplied by a variety of fuel sources. Generators are positioned on the merit order according to their variable costs, flexibility and contractual obligations. Some types of generation occupy the lower 'base-load' element of the merit order.

Nuclear and gas are used as base load on the basis of both cost and contractual obligations.

Hydro, where available to the utility, is used to supplement the base load. Generation from hydro plant is small in the UK context, although much more significant in Scotland. Hydro production has flexible and inflexible components which are a function of

Opportunities and Advances in International Power Generation, 18–20th March 1996, Conference Publication No. 419, © IEE, 1996

recent weather patterns. This determines its position in the merit order.

The difference, if any, between demand and generation is met by the combustion of black fuels. Coal is becoming increasingly mid-merit in the UK following the expansion of less flexible gas generation.

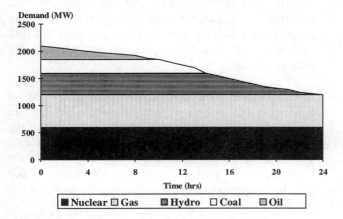

Figure 2: Typical Load Duration Curve

Competition in UK electricity generation has concentrated in the base load component of the generating market since deregulation in 1990. This is largely due to new entrants building modern high efficiency gas fired power stations. The survival of these new entrants will be critically dependent on achieving the lowest cost base.

4. CONTRACTS

Each fuel type used for electricity generation is purchased in a different way, allowing maximum flexibility with regard to security of fuel supply. Power station fuel can be purchased in a number of ways:

i.   'Take-or-Pay' Contracts: An Electricity Utility can agree with any supplier to take a minimum amount of fuel at an agreed rate over a particular period. They must 'take-or-pay' for this energy in accordance with the contract.[3] These contracts are generally related to energy which is not stored. e.g. gas.

ii.   Standard Contracts: Here a supplier tenders for a contract to supply a specified amount of fuel within a period, from a few months up to several years. Such contracts usually apply to fuels which will be stored. e.g. coal and oil.

iii. Spot Market: These are short term purchase agreements where energy is supplied quickly and at a price close to marginal cost.

Figure 3 shows how, for a typical multi-fuel generator, the contracts would fit together to ensure flexibility in fuel supply.

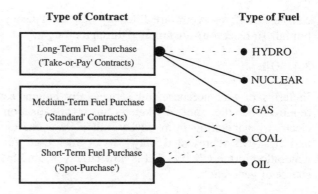

Figure 3: Fuel purchase contract types according to fuel.

5. PORTFOLIO MANAGEMENT

In managing fuel purchase, decisions must be taken on types, timescales for purchase and quantities in which to buy. Operating a portfolio of fuel types and suppliers offers the Generator a means to achieve more flexibility and, hence, lower cost solutions. This section considers different fuels from the point of view of associated contract types and flexibility in use.

5.1. Nuclear

Although, for technical reasons nuclear generation may lack the ability to respond rapid to changes in system demand, it can respond to market conditions given time. At the time of writing nuclear Generators have priority over other forms of generation when selling electricity into the UK Electricity Market. They have 'take-or-pay' contracts which allows them guaranteed sales of all the electricity they produce.

5.2. Gas

In the UK between financial years 1992/93 and 1993/94 electricity generation from gas increased by more than 700% [4]. At present 13% of the UK's total electricity generation comes from gas [5].

Electricity generation from gas tends to be inflexible due to contractual obligations to their fuel suppliers. Neverless, the commercial decision to 'turn-down' gas and pay a penalty to the supplier may be taken, despite the economic penalties which this attracts. Utilities can also increase their options through operating several gas fired power stations and by having direct involvement in the gas market.

5.3. Hydroelectricity

Historical data on run-off within catchment areas is used to forecast generation of electricity from hydroelectric plants. Since rainfall is not controllable, water appears at first sight to be inflexible. Pondage normally allows some degree of flexibility although this can be lost if reservoirs are filled by a spell of heavy rainfall, or reservoirs are emptied through drought.

5.4. Oil

In terms of fuel purchase, oil is generally bought as required on the spot market. Since privatisation electricity generation from oil has fallen by a quarter and by 1993 represented only 8% of fuel used for generation in the UK [6]. Oil is now largely a stand-by source of generation.

5.5 Coal

As previously mentioned, coal is increasingly taking a mid-merit role in the UK electricity generating industry. It now tends to occupy the position between less flexible generation and demand.

Because of its mid-merit position, black fuel purchase inherits the planning variance from other sources of generation and demand for electricity.

Experience gained from dealing with the practicalities of coal purchasing, for example transport and processing, suggest a lead time of approximately three months from initially identifying a need to purchase coal to first deliveries. Generators must therefore ensure sufficient coal is stocked to cover uncertainty in consumption over this lead time.

Generators operate a portfolio of coal supply contracts over a range of time periods, from a matter of months to a number of years. Usually only a proportion is committed to longer term arrangements, the rest is left to shorter-term (less than one year) purchases.

Reasons for purchasing coal under longer-term contracts:

i.   It creates supply by giving coal producers the stability and confidence to invest in new mines.

ii. Mines usually have some capacity to increase production over the base-supply contracts. Generators can therefore purchase additional tonnages at close to the marginal cost of producing the coal. This lowers the average price.

iii. Risk management technique of 'locking-in' some prices for a period ahead.

Apart from the price benefit, some of the coal demand is left to short-term purchases because of uncertainty in coal requirements in the developing market.

The balance between the volume of coal committed to long-term contracts and short-term 'spot' purchases is set by the degree of certainty in planning and the economic incentive of potential marginal cost supplies. Utilities must therefore rationalise the benefits of security of fuel supply with the uncertainty associated with the changing industry.

6. RESULTS

While it is possible for utilities to ignore the uncertainties associated with fuel purchase, simply because the losses are not apparent, this does not take advantage of positive opportunities which may otherwise be missed. [7]

In the previous discussion of portfolio management, it was stated that uncertainty is an important influence in the organisation of contracts. The formalisation of the fuel purchasing decision process and its related uncertainties is therefore central. This study will aid the development of decision reduction techniques which will be incorporated into interactive decision support software.

All forms of generation have related production difficulties and unplanned outages, even base load. For each individual fuel type used by the Generator in this study the difference between its planned use and its eventual use over the same period have been calculated. It is this set of differences and their causes that this work aims to analyse and so develop techniques for optimum fuel purchasing.

By comparing the actual generation from each source with the planned usage at the start of the selected period the major influences on coal use can be found. Further analysis of the figures gives an indication of what the differences between planned and actual generation mean in terms of black fuel stock required to meet a shortfall from another source, or extra reserves left due to an unexpected increase in output from elsewhere.

If gas and nuclear energy are assumed to be base-load then the difference between planned and actual generation can be analysed; thus the causes and effects of these differences can be examined.

A graph of the differences between actual output and planned output by fuel source is given in Figure 4. It can be seen from this that whenever the actual output from gas or nuclear is less than the planned schedule (a negative figure on the graph) the deficit in generating availability is made up by extra generation from black fuel. It can also be seen that whenever demand is higher than expected there is increased black fuel burn. Since black fuel burning makes up the difference between the 'Take-or-Pay' electricity supply (from gas and nuclear) and the demand it can be expected to reflect any variances between plan and actual use.

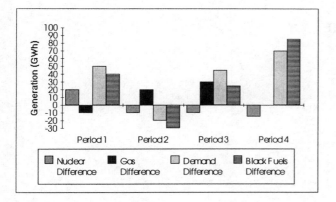

Figure 4: Graph of Differences Between Actual and Planned Generation and Demand.

For the example utility described in Figure 4 above it can be seen that given the two 'take-or-pay' contracts and the relative reliability of those sources, black-fuel generation mainly reflects variance in the demand. The following 'first-stage' analysis might be made:

Period 1.
- **Nuclear** - In this period there has been an unplanned increase in generation from nuclear fuel.
- **Gas** - Generation from gas is lower in this period than had been planned.
- **Demand** - Demand is greater than had been expected in this period, this may be due to cooler weather than was forecast.
- **Coal** - Coal generation has had to be increased in this period to counter the effects of variance in output from other fuels and demand.

Period 2.
- **Nuclear** - Nuclear generation has been less than planned in this period.
- **Gas** - Gas has shown an increase in output in this period.
- **Demand** - Reduced demand could be attributed to comparatively warm weather, reducing requirements for heating.
- **Coal** - The reduction in demand and the increase in gas generation has only been partly offset by the reduction in nuclear generation, hence coal has not significantly changed.

Period 3.
- **Nuclear** - Nuclear generation has been less than was planned for this period.
- **Gas** - Gas has shown a greater increase in output in this period than the last one.
- **Demand** - For this period demand has been higher than expected.
- **Coal** - The combination of differences in supply and demand has meant that more coal has been used than was expected.

Period 4.
- **Nuclear** - Nuclear generation has again been below expected levels.
- **Gas** - Gas output has been exactly as expected.
- **Demand** - Large increase in demand possibly due to low temperatures and increased requirements for heating.
- **Coal** - Reduced nuclear output and high demand have combined to increase coal generation.

The consequences of large variations in output from other sources, and the fluctuations which can occur in demand, can result in substantial variation in coal use.

7. CONCLUSIONS

Initial work in this field made it apparent that many of the factors influencing fuel purchase were little studied or understood, and that their interaction was not documented. Commercial imperatives in the privatised Electricity Supply Industry may now bring added weight to maximising fuel purchase effectiveness.

In the future, as the merit order for generation changes and black fuel use becomes more marginal new measures will be required to cope with the uncertainties created by the variance of coal generation. As coal generation becomes more marginal variance will become more significant relative to overall generation and coal demand will begin to appear more volatile.

Each Generator must take advantage of every opportunity available to improve its competitiveness. If the efficiency of fuel procurement is to be maximised in the long-term the factors which cause most variation in its use will have to be taken into account. The work being undertaken here will culminate in the formalisation of the fuel purchasing decision process and its related uncertainties. The project objective is to provide a valuable DSS tool for Generators in the ESI.

REFERENCES

1. Energy Trends, Department of Trade and Industry, July 1995.
2. RILEY, S. and TAYLOR, M.: 'The UK Generation Market', Power Engineering Journal, December 1993.
3. LEE, F.: 'The Coordination of Multiple Constrained Fuels', IEEE Transactions on Power Systems, Vol.6, No.2, May 1991.
4. Electricity Association, 'UK Electricity '94', p.60.
5. Energy Trends, Department of Trade and Industry, September 1995
6. Electricity Association, 'UK Electricity '94', p.27.
7. FOLEY, J.: 'Integrating the Corporate Attitude to Risk', Energy Price Risk Management Conference, 5th-8th October 1993

SCHEDULING CCGTs IN THE ELECTRICITY POOL

A P Birch[*]
M. M. Smith[**]
C. S. Özveren[***]

[*] Scottish Hydro Electric
[**] M M Smith Consultants
[***]University of Abertay

ABSTRACT

This paper presents a review of the scheduling problem of non traditional generators in England and Wales. It discusses the problems that have been faced, since the Privatisation of the Industry in England and Wales, by the new tranche of generation which is unable to conform to the classical bidding structure of the original CEGB plant. The more recent capacity comprises combined gas turbines and steam turbines. Such units which may operate as a single unit, or combination other like units and/or a steam turbine. It is not possible to create a bid pattern for these interlinked modes of operation which fully reflects the cost structure which is represented by the existing three point linear fit.

This paper reviews different methodologies that were considered for the scheduling of CCGTs to account for their own unique constraint characteristics and modes of operation. It reports in detail the method that was favoured by a number of the actual CCGT operators, from a technical and economic view point.

INTRODUCTION

Since the privatisation of the Electricity Supply Industry (ESI) in the UK, scheduling of the generators has become a matter of critical financial importance for power station operators. The National Grid Company (NGC), as the Grid Operator (GO), performs the scheduling and real-time despatch of Centrally Dispatched generators to meet National Demand. NGC is required by its license conditions to ensure that the scheduling process should be fair and reasonable. The schedule produced is used for two different purposes, firstly, it is used as the basis for the derivation of the system marginal price (used in the determination of the price that the customers will pay) and secondly, dayahead, it is used in real time system operation, to decide upon the actual output of the power stations on the system.

When privatising the electricity industry, one of the Governments requirements was that the privatised industry should still have an incentive to minimise the overall National production cost. This was achieved in two steps. Firstly, the generators were encouraged to offer their capacity to the central dispatch mechanism based on their cost structures, by giving them an incentive to bid prices close to, or at cost [1]. Secondly, NGC was required to minimise the National production cost based on these generator bid prices.

SCHEDULING

The purpose of scheduling generation (gensets) is to meet consumer demand at minimum total production cost, selecting appropriate generator start and stop times and associated output levels, while obeying all physical constraints of the generator and transmission system. This aim is achieved by creating simple algorithmic models of the generator cost structures and physical limitations (local constraints); and subsequently using an optimisation algorithm to choose between these generator models, deciding start/stop times and outputs so that demand and reserve are met and transmission constraints (ie global constraints) are obeyed. The ideal choice is the minimum total cost schedule which is feasible (ie obeys all constraints).

The existing scheduling tool minimises production cost based on single unit generators with linear Willans Lines with many assumptions based on the central organisation of production typical to a Nationalised industry. It has been decided to replaces this scheduler with a more modern tool [2]. It is expected that this scheduler will co-ordinate individual generation optimisation modules on a values basis using Lagrangian Relaxation. The individual generators modules will balance cost and values using dynamic programming. The final generators outputs will be optimised via a series of linear programmes. It is not necessary that the generators optimisation modules be identical, it is perfectly reasonable for the optimisation modules to be tuned to the cost and constraints characteristics of different types of generation.

The present and proposed scheduling tools both assume that all types of generator can be modeled appropriately by one cost and constraint structure, which is based on the concept of a single unit generator with costs and constraints based on the CEGB plant which was coal and oil fired thermal plant that offered its capacity to the central scheduling mechanism using a three point linear structure [3]. The individual generators will have differing values of the various cost and constraint parameters, but are restricted to remaining within the overall structure. There is no theoretical reason why the structure of the Generator models should be identical., but where Generator models have vastly differing structures then it is likely that it will be difficult to design an overall optimisation algorithm.

REQUIREMENT TO MODEL CCGTs

The physical constraints of the output of a combined cycle gas turbines (CCGT) plant are markedly different from that of conventional generating stations. The existing bidding procedures place a constraint on this concept as there are a number of conflicts which disadvantage CCGT operators.

The future capacity provided by the new CCGT stations [4] is becoming significant. In the year 1997/98, in which the new scheduling tool is hoped to be operational, only 39.5% of the generating plan will be able to match their operating characteristics to the existing bid structure compared with 23.7% of generating plant will be of the CCGT type.

BIDDING STRUCTURE

The bidding structure used by the scheduling mechanism (Generator Ordering and Loading Program - GOAL) is that of the Willans Line. Generators may bid a start-up cost, a no load cost and a maximum of three incremental costs (with associated elbow points). It is assumed that all generators can be modeled appropriately by one cost and constraint structure, which is based on the classical CEGB plant characteristics of coal and oil fired thermal units. The individual generators have differing values of the various cost and constraint parameters, but are restricted to remaining within the overall structure.

The new generation of CCGTs have cost structures that are significantly different from historic coal or oil fired thermal plant. The primary difference is that the CCGT is a module comprising an interlinked set of gas turbine (GT) and steam turbine (ST) generating units. There are cost inter-linages between the units and also a variety of time stagger constraints between the units. When considering optimum operation of the total module there are clear relationships between the units output. If additionally, frequently response and reserve capability are considered, sub-optimal operation is required and these relationships are then different.

CCGT CONSTRAINTS CHARACTERISTICS

The operation of a CCGT plant is constrained by a variety of operational and commercial considerations. These include:
Gas supply agreements - determines ramp rates, notice periods and pipeline storage.
Commercial considerations - PPA and CfD, gas take or pay and operating configuration costs.
Sensitivity of plant output to ambient conditions.
On-site load - electrical and steam.

CCGT units differ from the more conventional thermal units by a number of factors, that include GT start-up profiles being the same whether the unit is hot or cold.

The module start-up time and profile is dependent upon the steam quality, temperature of the steam turbine and configuration of the boiler and steam bypass systems.

SCHEDULING APPROACHES

There has been much debate concerning the area of scheduling CCGTs, but the three practical approaches that were considered [5] are scheduling by: physical units, by unit combination and by MW ranges.

Physical Units
In this approach the units are modeled individually and the scheduler puts them together in a building block style. The modeling of the individual unit constraints is straight forward, but the scheduler must optimise joint GT and ST outputs as the steam unit output is wholly dependent on the GT(s) operation. It must be recognised that the ST production cost is dependent upon the number of GTs in operation at any time.

Unit Combination
The scheduler models an allowable combination of units (both GTs and STs) as a single entity. These combinations are then swapped during the day as the interaction of constraints permit. When a single combination is running, its output optimisation is

straight forward, as it offers a single cost curve which incorporates both ST and GT outputs. However, the physical constraints, such as the ST minimum on time, are difficult to model.

MW Ranges

This approach is essentially a restriction to the scheduling by unit combination, whereby permitted unit combinations do not have overlapping output, or ranges. The station operators do not state which combination of units, applies to each range of MWs, provided that all ranges offered may physically be achieved. This method is simpler to understand and perhaps to implement, as it has no requirement to monitor which units are operating to define the combination. For the reasons of simplicity and the operational freedom that this approach provides for the station operators, this was viewed as the preferred method.

COMBINATIONS AND RANGES

In a centrally planned power system, the GO will instruct the power station operator on how to run the individual generation units (gensets) and is assumed to obey such instructions. This is possible as the GO is assumed to know exactly the full cost, efficiency and constraint information pertaining to each genset, however, in practice these assumptions are not always correct. Hence in this environment, the GO considers all possible CCGT combinations if completely feasible and appropriate.

In a "pooled" environment, where the genset operator is being paid according to bid prices and MW output, the combination method presents difficulties. The primary difficulty is for the GO and the Pool to determine which combination is running at a particular output MW level. If they are unable to distinguish, the environment is given to gaming by the genset operator. Conversely, if the GO tries to dispatch, physical combinations then physical genset switching information should be incorporated within the Pool, which is not robust and this adds enormous extra complication. It is worth noting that the Pool in England and Wales does not have any concept of ON and OFF, it only has the idea of 0 MW measured output and 1+ MW output.

The ranges method accepts the constraint that only MW output is meaningful and that physical combinations providing the output is irrelevant for payment and scheduling purposes (although it is still important for fault levels etc.). Pragmatic operation of the ST improves this, but has minimum running constraints, hence for the base load operation the

sequence would be GT, ST, GT and for peaking GT, GT and ST.

Algorithmic Problems The primary concern in applying the ranges method with a schedule that optimises (gensets output) utilising dynamic programming is the treatment of physical unit minimum on and off time constraints (which also applies, perhaps to a greater extent, in any practical application of the combinations method). It was decided that the pragmatic approach was to treat the condition of minimum off and minimum on differently and to recognise that the major constrains are those of the ST.

Some of the formulations of this problem incorporate the concept of warm up delay between the start of the first GT and the ST. In this approach this is dealt with by utilising RUR. (which perhaps are very low)

Minimum on time (MOT) The ST runs within the run time of the unit as a whole it is not possible to control its minimum running by use of a module MOT. It should also be noted that applying a MOT within the range (or the combination in the case of the combination method), unduly constrains the solution.

It is necessary therefore to utilise the concept of minimum time between range up and down. This means in the simple example, the true constraints are applied to the summated time in R2 and R3. This causes a conceptual conflict with the DP process which prefers naturally to forget information concerning previous states. This conflict is perhaps best cured by utilising the concept of state and sub state. This would means labelling the range change which incorporated the ST start/stop as the primary transition, in which case R0 and R1 are the low states and R2 and R3 are the high states.

Minimum off time (MOFF). This could be treated in a like manner to the MOT. Alternatively, it can be noted that the MOFF of the unit is constrained within the MOFF of the ST, provided the module is required to shut down whenever the ST does so.

In the simple example, this means that whenever the module ranges down from R2, it must shut down completely before it can again range up into R2. In figure 4 this is achieved by having 2 separate states in which the ST of fully shut down with the GT running. Note with a ST MOFF of 3 hours the module MOFF would only need to be 2 hours as the first GT would need to shut down and start up in between.

State Space and Transitions Figures 1 to 4 show a comparison of the relative sizes of the genset local solution problems using DP for a classic generator and a CCGT. The run time for a DP is dominated by the number of states and transitions. Figure 1 shows a classic genset with 20 states and 22 transitions.

Figure 2 shows the on/off diagram for a 2+1 CCGT with the base load running ranges shown in bold.

Figure 3 shows the combinations appropriate to the ranges offered and is of course only a part of the full combinations state space diagram. It also contains 3 other layers representing the full set of paths around the on/off diagram (as in figure 2). For the combinations method the number of states are 18 and number of transitions are 38.

Figure 4 gives the state space diagram described below and has 19 state spaces and 34 transitions.

RANGES METHOD

This method does not allow a full description of the physical process as was possible in the first two methods. In both of the previous methods, a single output level could have be achieved by more than one set of units with the associated cost. The output could be modeled as coming from different sources. In the ranges method all the output levels are unique and associated with only one combination. Clearly, there is some sacrifice made here, as the station operator must lose the ability to fully model one or more physically feasible unit combinations.

The ranges method is a pragmatic restriction of the combination method in which each output level of the module can only be represented by one combination. This, however, prevents the proposed approach from fully modeling the full operating conditions of a CCGT. In theory it is possible that a global optimal solution may not be found. However, in practice this restriction will not be so great as the operator will be able to choose which physically possible combinations they offer to be associated with each range of output.

OPERATION AND SCHEDULING OF CCGTs

The approach described in this paper will allow each sub-module to be modeled as a single entity. These sub-modules may then be swapped by the operator, during the day, as their interaction constraints permit. While it is running, the output optimisation of each sub-module is straightforward, as it offers a single cost curve which incorporates both ST and GT outputs.

The operator has the freedom to chooses which of the physically available sub-modules to offer to the system. Each sub-module is offered (and delineated) by a

non-overlapping range of MW module output. These Ranges are then ordered and will abut (ie: the lowest output of Range 2 will be one MW greater than the highest output from Range 1).

Each Range will be associated with a sub-module, which will be the operator's preferred method of producing that output in steady state (although plant failure and restrictions during run-up and run-down may cause the operator to produce that output from a different sub-module, at a cost penalty).

CONCLUSION

The preferred method of modeling generators gives a unique cost - power curve which is represented by a series of genset configurations, all of which are mutually exclusive (ie only one genset configuration is allowed to run at any one time, and the configurations must be loaded and de-loaded in a set sequence). The model shall be capable of including start-up costs for each configuration element or module.

The ranges method of offering CCGTs has been selected by the independent operators as it provides the most flexible approach. It enables the operator to offer the module in a manner that is the most appropriate for the days circumstances, accounting for local operating constraints, both physical and economic. The suggested approach may restrict the ability of the scheduler to select the theoretical overall optimum selection of generators for the whole system, but does provide the individual operators with much needed flexibility and control. It has the added benefit that the proposed method requires the least number of changes to other existing systems and, both proposed scheduling algorithms and settlements systems.

REFERENCES

1. Birch, A. P., Smith, M. M., and Özveren, C. S., "A Review of the Electricity Supply Industry in Britain", MELECON'94, 7th Med. Elect. Conference, Antalya, Turkey, April 12-14, 1994.
2. National Grid Company, "GOAL Replacement Joint User Requirement Specification", Issue 1 Definitive, 1 July 1993, Pool Members and NGC Classified.
3. The England and Wales Pool, "The Pooling and Settlements Agreement", Revised 16 November 1992.
4. National Grid Company, "Seven Year Statement", Sixth Edition, March 1995.
5. Birch, A. P., Smith, M. M., and Moody, P.,"GOAL Replacement Project JURS for scheduling CCGTs", Issue 2, Draft 2, 25 March 1994, Pool Members Classified.

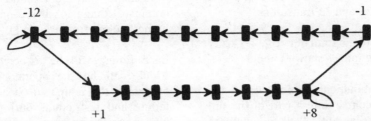

Figure 1: State space diagram for a classical genset which has a minimum on time of 4 hours and a minimum off time of 6 hours, assuming a half hour scheduling interval. Stable states are indicated by the loops.

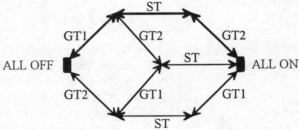

Figure 2: Start and stop diagram indicating possible pathways for a typical CCGT module. The bold path is the one the genset operator would normally prefer and underlies figures 3 and 4.

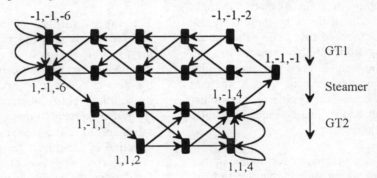

Figure 3: Partial state space diagram showing all combinations of states and transitions for the CCGT in figure 2, which has a minimum on time of 2 hours and a minimum off time of 3 hours for the steamer. The Gts are able to start and stop in half an hour. The full diagram has three more layers.

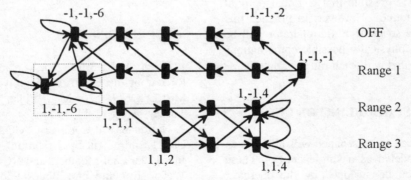

Figure 4: As figure 3 but with the additional constraint imposed by the Ranges method. Note that this is the full state space diagram. Minimum off time is defined for the module as a whole and is two hours imposed by that of the steamer, this creates the need for another node as the module must fully shut down before the steamer can restart. The steamers minimum on time is applied as a constraint on the summated run time in ranges 2 and 3.

OFFSHORE POWER GENERATION - LIMITED LIFE EXPECTANCY

I D Stewart

BP Exploration Operating Company Ltd

INTRODUCTION

The offshore oil industry has been well established on the United Kingdom Continental Shelf for some 20 years. In that time some hundreds of Megawatts of power generation has been installed on oil and gas production platforms. The size of power generators has gradually increased as has the electrical load on the platforms. As the production declines and new frontier areas are opened up, what further opportunities are there for the power generation industry? This paper looks at how the development of platform power system design has changed over the years to the present day and presents one future scenario that may not include any power generation on the platform itself.

ELECTRICAL SYSTEM DEVELOPMENT

First Designs

The first platform electrical power system designs had the larger equipment, such as gas compressors and main oil line pumps largely powered direct from gas turbines with the electric drives generally being of less

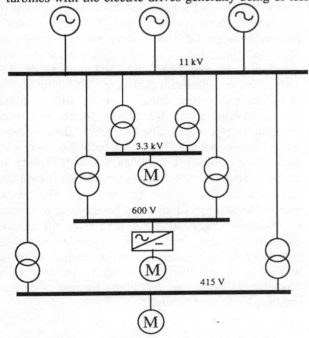

Figure 1: Early platform electrical system design

than 1 MW. A typical distribution system is shown in Figure 1. There are 3 by 50 % rated generators connected at 11 kV for the power system. Motors are connected at 3.3 kV (typically sea water lift pumps), 600 Volt (dc drilling motors) and 415 Volts (motors up to about 150 kW rating). There are no motors connected at 11 kV as the larger loads are gas turbine driven. The total power system load was about 6 MW

Designs of some 15 Years Ago

The power system of a larger oil production platform is shown in Figure 2. The voltage range is the same but the total platform load is about 40 MW. There are 3 by 50% rated generators with actual ratings increased to about 25 MW. The compressors and other large pumps are electric driven being connected to the 11 kV busbar and with ratings up to about 8 MW. Power system design has to cater for starting of such large motors. Conventionally the systems were designed for direct on line starting. This has often resulted in the need for close co-operation with the alternator manufacturer in the control of the design to meet closer tolerances of the sub-transient and transient reactance than would normally be permitted in the standards. Starting large motors direct on line requires the transient reactance to be low but the limit of switchgear fault ratings require the sub-transient reactance to be high. The design of the two reactances of course is not independent. The feeders from the main 11 kV busbar are duplicate radial feeds via the transformers to lower voltages.

Present Day designs

The loads continue to grow with increasing size requirements for compressors and pumps. The total platform load is increasing towards 100 MW which requires either an increase in number of gas turbine generators or an increase in individual rating. In addition there is an increasing need to keep operating costs under control and therefore activities requiring significant maintenance are constantly reviewed to attempt to keep them to a minimum. Therefore some platform designs are now considering utilising a generation configuration of 2 by 50% or even 1 by 100% plus some small unit to provide more than just

Opportunities and Advances in International Power Generation, 18–20th March 1996,
Conference Publication No. 419, © IEE, 1996

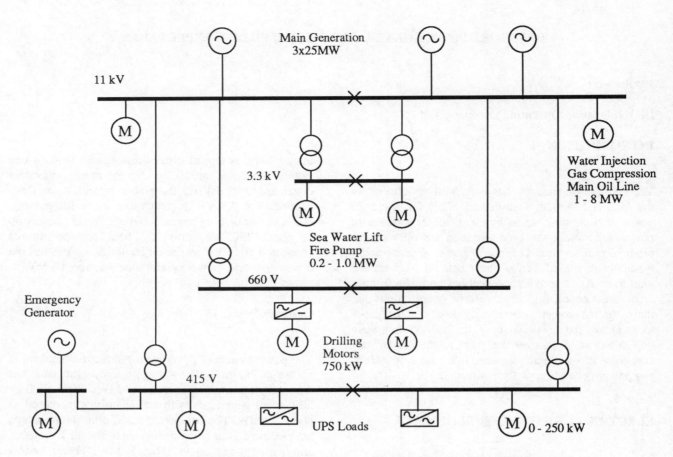

Figure 2 : Typical Electrical System Design of Large Oil Production Platform

emergency loads. Maintenance requirements are more often based upon performance and condition monitoring techniques plus the use of reliability centred maintenance as a tool to direct work at critical items. This could lead to reduction in manpower requirements to be based offshore which can have significant cost benefits. Although large gas turbine drives for gas compressors are still used the increase in number of turbines does increase maintenance costs. Some platforms are utilising large power electronic variable speed drives for gas compression duty and a typical system is shown in Figure 3.

When reduction in the number of generators is considered then the availability to the critical loads has to be considered. Some critical activities such as drilling will be carried out by jack-up rigs alongside the platform so that only small workover loads are likely. The need to provide assurance against loss of power to the drill string when it is critical is no longer a problem.

The high levels of power electronics means that harmonic distortions are significant and have be carefully designed. The use of 24 pulse convertors helps reduce the low orders of harmonics plus use of filters.

Harmonic distortion of the voltage waveform has been known for many years. The prime sources and effects have been presented elsewhere by Yacamini et al (1), Yacamini and Stewart (2) and Stewart (3). It is worth noting that on offshore isolated power systems that the level of distortions have reached levels well in excess of the normal limits given in the Electricity Council Recommendation G5/3. The limit would normally be 5% Total Harmonic Distortion (THD) whereas some installations have operated with THD in excess of 15% with as yet no significant problems. The main source of harmonic distortion comes from dc drilling drives and ac variable speed drives. The former give very transient disturbance whilst the latter gives a more steady state condition. The main problems seen with power generation have been limited to difficulties in synchronising generators where the voltage waveshape had additional voltage zero crossings and affected the relays. The closing of the circuit breaker was prevented so it was a safe situation but gave some operational difficulties at that time. Another problem related to failures of fuses in emergency light fittings occurring when drilling operations commenced. An input filter capacitance gave rise to high harmonic current resulting in the fuse failure.

Harmonics

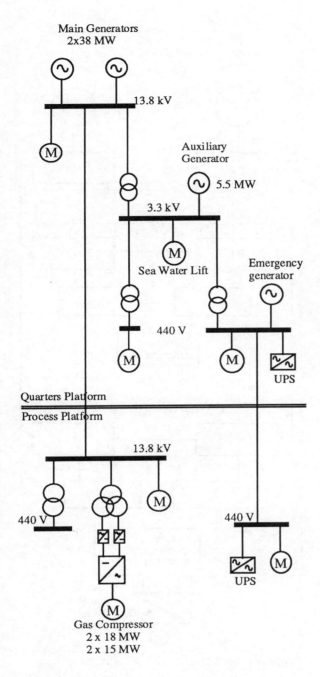

Figure 3 : Present Day Electrical System

FUTURE DEVELOPMENTS - NO GENERATION

New Frontier Areas

What could the future electrical power system designs look like? For the new frontier areas such as West of Shetland, the current trend is using Floating Production, Storage and Offshore loading (FPSO) with a vessel that is basically a ship with a production facility in the middle to cater for the field. Power generation and power distribution systems are therefor

generally conventional as they are required when the vessel is operated as a ship and has to comply with International Maritime Rules.

North Sea Developments

The Fields in the North Sea left to develop are far smaller than the initial large discoveries. New ways to make the development of these fields economic for Operators are constantly being sought. Operating costs are high and reduction in maintenance can be a significant factor. Typically the annual cost of maintenance for turbines on an offshore installation is about £1.5 to £2.0 million. As the field production declines the amount of available gas to burn as fuel also declines to a point where imported liquid fuel is needed. Where a field is close enough to an existing facility subsea power cables may be used provided the existing platform has spare capacity. Many developments will therefore have stand alone generation or be connected to existing fields.

Novel DC System

There have been many studies using conventional DC links to the shore, but because of the distances, voltages concerned , the overall benefit was not enough to make an operator choose this option.
This study therefore took a different approach to see what needed to be achieved to make dc distribution an alternative that could be considered.

The required aspects of such a scheme would be:-

Compactness. Conventional DC systems are physically large. Development in semiconductor technology could help to reduce this.

Connectivity. To offer economic solution and be advantageous then a number of offshore installations need to be connected together in such a way that they have minimal affect on each other.

Input and Output. To take advantage of offshore capacity already installed it would be beneficial to have some platforms feeding spare capacity into the network with other being loads only. This would reduce the demand from the beach.

Minimal weight. The cost of weight in an offshore installation is very significant.

Communication. By connecting platforms with power cables it is possible to install fibre optic cables within the power cable thus providing direct means of communication. In addition unmanned platforms could have cctv installation to aid operation and safety.

Block Diagram. A block diagram of a sort of scheme can be seen in Figure 4. The figure shows the scheme for an existing installation, which has a large number of small turbines (33), and a proposed new development.

Overall Benefit. The existing installation would have all turbines removed, electric drivers installed and eliminate the need to import liquid fuel. For the scheme outlined above it was calculated that a NPV saving of about £20 million was possible. The new development would have no main power generation, have variable speed drives connected direct to the dc supply. Three power generation units of about 40 MW would add an overall weight of about 670 Tonnes. Variable speed drives would add a further 280 Tonnes. For the new development the potential weight saving was estimated to be about 700 Tonnes, which in capital expenditure alone amounts to some £68 million, and overall an NPV saving of about £60 million was possible.

Drawbacks. Of course it is never as simple as portrayed in this paper. Firstly the technology has yet to be provided economically to achieve such scheme. Secondly for existing platforms, such a scheme requires a significant capital investment programme when conventionally no such expenditure is planned. Thirdly the infrastructure to provide such a scheme for a new development would have some degree of risk and operators are very sensitive to power loss when it affects production. Contracts for supply of gas are particularly sensitive and the risk difficult to estimate compared to conventional generation.
Finally in a large multinational company there are many schemes being proposed to secure part of a finite amount of funding and there has to be careful evaluation of the ones which have the best return against risk in a changing environment of technology and politics.

CONCLUSION

Power generation using gas turbine driven generators have been in service in the offshore industry for many years now and have given good service. The rating of generators is tending to increase due to the increase in use of power electronics. Future developments in power electronics in providing converter equipment that could replace generators totally has been found to be commercially attractive and may offer the offshore oil industry an alternative that could extend production well into the next century

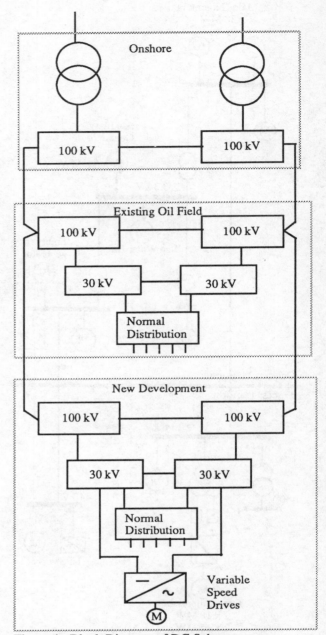

Figure 4 : Block Diagram of DC Scheme

REFERENCES

1. R Yacamini, L Hu and R A Fallaize, "Calculation of commutation spikes and harmonics on offshore platforms", IEE Proc Vol 137, Pt B, No 1, pp35-48
2. R Yacamini and I D Stewart, "Harmonic problems and calculations on offshore platforms", Trans IMarE, Vol 102, pp249-259.
3. I D Stewart, "Offshore Power System Harmonic Distortions - Problem, Nuisance or Fact of Life", ERA Conf Proc, Protecting Electrical Networks and Quality of Supply in a De-Regulated Industry, pp 5.2.1 - 5.2.11

DEVELOPMENT OF PRIVATIZED POWER INDUSTRY IN MALAYSIA

A Akhtar,   A Asuhaimi, H Shaibon   K L Lo

University Technology Malaysia, Malaysia   University of Strathclyde, Glasgow

Abstract. In the last few decades reliable supply of electric power has become a central issue in the development and economic prosperity of any country. Recent years have witnessed far reaching and radical developments taking place in the electric power industry. Among other developments, privatization trends in power generation are having significant implications on the planning and operation of electric utilities.

Over the last few years, Malaysia has experienced a rapid economic growth resulting in an ever increasing demand of electricity. This has prompted the Government of Malaysia to invite private power companies to build, own and operate power generation facilities to meet the growing needs of electric power in the country.

This paper provides an overview of the development of privatized power industry in Malaysia and the government efforts to liberalise the elecric power sector in order to create an environment of increased competition and efficiency. The paper presents a brief history of the private power generation in the country, and provides the information regarding technical, financial and commercial aspects of the power generation projects constructed in the private sector. Finally, the paper presents a case study on the implementation of the private power plant project, and various technical and operational issues confronted by the owners of the project.

INTRODUCTION AND BRIEF HISTORY

The privatized power generation in Malaysia dates back to early 1900 and at that time was largely associated with the tin mining industry. The notable pioneer private companies were Malacca Electric Light Co. Ltd. (1913), Huttenbachs Ltd. (1916), Perak River Hydro Electric Co. Ltd. (1926) and Kinta Electric Distribution Co. Ltd. (1928).

The   process   of   subsequent   integration   and consolidation of the private power producers saw the elimination of most of the private companies by the end of the second world war. In 1948 the only remaining private power company, namely Malacca Electric Light Company was also taken over by the Government [1].

In 1949, the Central Electricity Board (CEB) was established with the responsibility of providing electric power to the whole of Peninsular Malaysia except for the state of Perak which was continued to be supplied by Perak River Hydro Electric Company Ltd. and Kinta Electric Distribution Company Ltd.. In 1965, the Central Electricity Board was renamed as National Electricity Board (NEB).   In 1982 with the nationalization of   Perak River Hydro Electric Company and Kinta Electric Distribution Company, the private power generation came to an end in the country [2].

During the last decade the Government of Malaysia actively pursued the policy of privatization with the following main objectives:

- to relieve the financial and administrative burden of the government;
- to improve the efficiency and productivity; and
- to reduce the size and presence of the public sector in the economy.

As a first step towards the privatization of electric industry, NEB was corporatised in July 1990 and renamed as Tenaga Nasional Berhad (TNB, the English translation being National Power Company Ltd.). In February 1992, TNB became fully privatized through an offer for the sale of its share to the public and subsequent listing on the Kuala Lumpur Stock Exchange.

The successful privatization of TNB heralded a liberalization and deregulation of the electric supply industry. As the only electric utility in the country, TNB was still   operating as   a monopoly, and it was perceived that the electric supply industry could be made more efficient and cost effective by introducing competition. Therefore, in 1993 the Government of Malaysia to initiate the induction of private power sector in the country by awarding licences to Build, Operate and Own (BOO) the power plants [3].

With the liberalization of the electric supply industry, over 45 proposals to build, operate and own power generation plants were submitted by the prospective Independent Power Producers (IPP's) for assessment to TNB, Economic Planning Unit and the Ministry of Energy. The nascent nature of these proposals and the lack of a standard appraisal procedure forced the

Opportunities and Advances in International Power Generation, 18–20th March 1996,
Conference Publication No. 419, © IEE, 1996

Government of Malaysia to centralize all independent power producers applications under a special energy unit created within Economic Planning Unit.

In April 1993, the first private power generation licence to build, own and operate two combined cycle power plants with a total capacity of 1212 MW was granted to YTL Power Generation Sdn. Bhd. The finalisation of Power Purchase Agreement (PPA) between TNB and IPP's, among many other issues, was the foremost issue that took considerable time in the initial stages of the development of private power industry in Malaysia. However, PPA negotiations between TNB and IPP's were facilitated immensely by the adoption of a new model PPA drafted by Economic Planning Unit with the help of a Canadian consulting firm. Under this power purchase agreement electricity sale pricing and payments are based on separate capacity and energy rates. The power purchase agreements are supposed to be identical in respect of their contractual provisions, however, the tariff for electricity sale by each IPP differs due to the mode of operation and initial capital outlays of each power producer [3].

INDEPENDENT POWER PRODUCERS

At present, the total demand of electricity nationwide is approximately 6,400 MW. It is envisaged that by the year 2000 the electricity demand will reach 10,000 MW, and thereafter it is expected to grow at a rate of 10% per annum until the year 2020 [4].

To meet the rising demand of electric power in the country, the Government of Malaysia started to issue licences to Independent Power Producers (IPP's) in 1992. Altogether, the government has awarded licences to five IPP's. Three of these IPP's plants will operate as base load plants, while the remaining two will operate on the peaking mode.

When all five IPP power plants will become fully operational by the year 1998, they will provide about half of the country's new power capacity, and about 30% of the total output that will be generated in the country.

In order to provide a vivid picture of the developments that have taken place in the private power sector in the country, this section provides the brief history of the implementation of the private power generation projects alongwith the features, and various technical, financial and commercial aspects of these projects.

YTL Power Generation Company

YTL Power Generation Company was the first IPP in Malaysia to be granted a licence to generate electricity. It has built two combined cycle gas turbine plants with a combined capacity of 1212 MW. The turnkey construction contract was carried out by a joint venture between a local company and Siemens AG of Germany. Siemens took the primary responsibility for design., construction, installation and commissioning.

In this project the concept of joint venture has been extended to include the operation and maintenance of the power stations. YTL has joined forces with Siemens to form YTL Power Services Company Ltd. which will be responsible for the operation and maintenance of the power stations. This means that in line with the global developments in the private power industry and the rapid growth of Build, Operate and Own (BOO) and Build, Operate and Transfer (BOT) projects, the designers and construction contractors have a continuing interest and responsibility for the operation and maintenance of the power stations.

In late 1994 the commercial production of electricity at both the stations began; the initial operation mode being open cycle. The power station located at Johor Bahru has one generating block consisting of two gas turbines and one steam turbine, while the plant located at Paka has four gas turbines and two steam turbines. Liaison with the fuel suppliers (Petronas) and with the TNB National Load Dispatch Centre is handled by official procedures, scheduled notifications and close working contacts on day to day basis.

Powertek Company

In line with the Government policy to rapidly induce private power generation in the country, in July 1993 Powertek received the necessary approvals from the Economic Planning Unit to urgently proceed to build, own and operate one open cycle power plant in Malacca. Under this licence Powertek has been authorized to build, own and operate a gas fired, 4 x 110 MW open cycle power plant at Malacca. Operating initially on a peaking mode, Powertek was given the option to convert the plant to a combined cycle plant before the year 2000.

The turnkey contract to construct the 440 MW open cycle plant was awarded to the consortium of European Gas Turbines SA - Alsthom Export Malaysia Company and a local firm Zelleco. The cost of the plant is approximately US $ 248 million in total and comprises of four 110 MW gas turbine generators of the General Electric Frame 9E type supplied by European Gas Turbines SA of France. All four machines were commissioned in early 1995 using natural gas supplied by Petronas as the primary fuel with an option to use distillate as a standby fuel.

Port Dickson Power Company

Port Dickson Power (PDP) plant consists of four 110 MW each GE MS9001E gas turbines arranged in two blocks of two units each. PDP gas turbines were the first to be fired among the five IPP's. The first two gas turbine generators were fired in July 1994 and the remaining two in August 1994. Commercial operation of Block 1 (units 1 and 2) commenced in January 1995, while the commercial operation of Block 2 (Units 3 and 4) began in April 1995. As of July 1, 1995, the Port Dickson power plant has accumulated 967 fired hours of commercial operation [4]. The primary fuel is natural gas but the gas turbines are designed to run on light distillate also.

PDP project is a joint venture among four leading Malaysian companies: Sime Darby Co. owns 40%, Malaysian Resource Corporation 30%, Hypergantic 20% and Tenaga Nasional 10%. Under a 21 year power purchase agreement, the power plant will supply power to TNB on an "as needed" or peak basis [4].

The total cost of the power plant is M$ 700 million. The main contractors of the project were General Electric Co. of USA and GE Power Systems (Malaysia). General Electric USA provided the main equipment and the engineering leadership, while GE Power Systems (Malaysia) was responsible for project management, all in-country activities for plant construction, sub-contractor management, and local procurement. Black & Veatch International were appointed as the consultant (owner's engineer) of the project. A seven year contract for plant operation and maintenance was awarded to ESB International of Ireland.

Under phase 1, Port Dickson power plant will operate as a open cycle peaking facility generating 440 MW. Under phase 2, depending on the feasibility, the plant will be converted to combined cycle operation adding another 200 MW of electric power capacity.

Genting Sanyen Power Company

Genting Sanyen Power Co. obtained the licence to construct 720 MW power plant in July 1993 from the government. Soon after obtaining the licence, the Genting Sanyen entered into an agreement with Petronas Gas Co. for the gas purchase contract and with TNB for the power purchase agreement.

Subsequently, Asea Brown Boveri was appointed as a turn-key contractor of the project. The total cost of the project was M$ 1.8 billion. Malayan Banking Berhad provided the finance to the tune of M$ 1.0 billion, while the balance of M$ 0.8 billion was made available from the company's own resources.

The Genting Sanyen power plant is a combined - cycle cogeneration facility. It consists of two GE Frame 5 gas turbines and three ABB 13E2 gas turbines and one steam turbine. Initial operation mode is open cycle, and it is envisaged that by early 1996, the plant will be converted to combined cycle operating mode thus generating its full capacity of 720 MW.

Segari Energy Ventures

Segari Energy Ventures (SEV) Lumut power plant with a total capacity of 1303 MW is the largest combined cycle power plant by an independent power producer in Malaysia. Taking into account the scope of the project and huge financing requirements of M$ 4.0 billion for the project, SEV was able to obtain support from a number of organizations. After many feasibility studies and negotiations SEV secured M$ 3.0 billion in the form of loan and bond facilities making it the biggest project financing in the country provided to independent power producers. A consortium of Bank Bumiputra and Maybank will provide M$ 1.5 billion in the form of floating rate loans, while M$ 1.5 billion in the form of fixed rate bonds will be arranged by Employees Provident Fund (EPF). The remaining M$ 1.0 billion will be generated from shareholder's equity and internally generated funds.

For the construction of the Lumut Power Plant, SEV has awarded a lump-sum fixed price, turnkey engineering, procurement and construction contract to Asea Brown Boveri (ABB). As per terms of the contract ABB guarantees electrical output, heat rate, emission levels and timely completion of the project and interconnection facilities.

The power plant will comprise of two identical 651.5 MW blocks each consisting of three ABB 13E2 gas turbines, three heat recovery steam generators, one steam turbine, four electrical generators, ancillary plant and equipment, interconnection facilities and common buildings.

The first block of 651.5 MW of the 1303 MW plant is scheduled to be completed by July 1996 while the second block is scheduled to become operational by July 1997. The engineering, procurement and construction contract provides for the payment of appropriate liquidated damages for failure to achieve the stipulated dates of completion.

The Lumut Power Plant is designed to run on natural gas or distillate oil, with natural gas being the primary fuel. SEV has finalized a gas sales and purchase agreement with Petronas who will also construct a 17 km pipe line for providing gas to the power plant.

The total net electricity output of the power plant will be sold to Tenaga Nasional Berhad (TNB). The power purchase agreement has incorporated a tariff structure which provides for a fixed capacity payment and a variable energy payment. The power purchase agreement is for 21 years from the date of commercial operation of the power plant.

As regards the operation and maintenance of the power plant, SEV has signed an Operation and Maintenance (O & M) agreement with Teknik Janakuasa Sdn Bhd (TJSB). However, TJSB has entered into an agreement with Black and Veatch International and ABB for technical services and consultancy assistance. The contract between SEV and TJSB is for a period of 15 years from the date of commercial operation of Block 1. The operation and maintenance contract covers the direct cost of operation and maintenance plus a fee for achieving specified performance levels.

CASE STUDY OF POWERTEK'S POWER PLANT

In early 1993 Powertek submitted its technical and financial proposal for the construction of a fast-track gas turbine power plant in Malacca. In July 1993, Government of Malaysia gave its formal approval and awarded the licence to Powertek to build, operate, and own an open cycle power plant at Malacca with the option to convert the plant to combined cycle by the year 2000. The bids were invited in December 1993, and following suppliers submitted their bids: General Electric of USA, Marubeni Corporation of Japan and European Gas Turbines (EGT)/Alsthom/Zelleco consortium. Tender evaluation was based on two envelope system whereby technical proposals were evaluated before the opening of the financial envelopes. The technical bid of Marubeni Corporation was non-compliant and was not considered further. The financial bids of GE of USA and EGT/Alsthom/Zelleco consortium were evaluated. Subsequently in January, 1994 the Engineering, Procurement and Construction (EPC) contract was awarded to the consortium of EGT/Alsthom/Zelleco.

The major arrangements for the EPC contract were that EGT will supply the main gas turbines, GEC Alsthom will provide the generators and balance of plant, and Zelleco will be responsible for the civil works and building services. For the 275 KV switchyard Ceglec of France was selected as the subcontractor, while Arab-Malaysian -Best and Crompton Co. were selected as a subcontractor for the transmission line.

Fuel Supply Arrangement

The Powertek's power plant is designed to run on natural gas with distillate as a standby fuel. The gas supply agreement with Petronas was signed in March 1994 after a period of three months negotiations. According to this agreement Powertek will bear the cost of construction of the 20 km gas line required to connect the main gas transmission line to the power station. Being a peaking plant, there will be no minimum gas offtake obligations by Powertek. For the standby and interim fuel Powertek has constructed fuel tanks with a total capacity of 15 million litres.

Project Management and Engineering

For the implementation of the power plant project, Powertek appointed an established local consulting firm KTA Tenaga as their technical advisers. KTA Tenaga were not only responsible for the engineering aspects of the project but also advised on the other aspects such as, financial matters, insurances, and operation and maintenance of the power plant.

As it was a fast track project, the main focus was on quality control, scheduling and cost management. In order to expedite the implementation of the project the EPC contractor's drawings, proposals, procedures and quality plans were reviewed at site. Critical path method was used to schedule and monitor the construction activities. In addition, fortnightly meetings were also held with the contractors to closely monitor and control the implementation of the project.

As regards the construction of the 275 KV grid station and transmission lines, close consultations were made with TNB on the project management and engineering aspects. Wayleave acquisition for transmission lines is normally a major hurdle in the timely completion of the project. To overcome potential problems in this respect, TNB applied on behalf of the Powertek and the State Government of Malacca also provided valuable assistance in acquiring the necessary wayleave acquisition. The operation and maintenance of the grid stations and transmission lines were handed over to TNB after their commissioning.

For the EPC contract, the original insurance requirements were limited to basic insurance policies applicable to such contracts, and these were the Contractor's All Risk (CAR) policy as well as insurance for marine cargo. However, as the financing of the project was on a limited/non-recourse basis, it was considered desirable to incorporate the Delay in Start Up (DSU) or Advance Lost of Profit (ALOP) covers. Finally it was decided to provide ALOP cover to both the marine cargo and Contractor's All Risk (CAR) policies. These additional covers were used as a tool to protect Powertek and its bankers from losses arising from force-majeure events.

For this project, the Government of Malaysia gave tax concessions on import duties and sales tax in respect of imported plant and locally manufactured equipment.

These concessions were in the form of sales tax and customs duty deferments.

Operation and Maintenance

For the operation and maintenance (O & M) of the power plant, Powertek invited tenders from established plant operators using a management type of contract. This requires the operation and maintenance contractor to be responsible for the operation and maintenance of the plant, and to provide key managers to direct, supervise and train local staff.

Five international bidders submitted their proposals and the following three were shortlisted based on the compliance with the tender criteria:

- PowerGen Plc of UK;
- IVO Generation Services (UK) Ltd. (Subsidiary of Imatran Voima Oy of Finland); and
- Electricity Supply Board International (ESBI) of Ireland.

Post tender clarification meetings were held with each prospective operation and maintenance contractor to discuss and finalise the commercial and legal aspects of the O & M agreement. Subsequently, IVO Generation Services were awarded the O & M contract. According to the terms of the contract, IVO has provided Station Manager, Operation Manager, Maintenance Manager and the Control/Instrumentation Engineer for an initial period of five years. In line with the IVO's recommendations, the total staff strength of the 440 MW open cycle power plant is 38.

Future Plans

At present Powertek's open -cycle gas turbine power plant is operating on the peaking mode. However, provisions have been made to convert the power plant into two combined cycle blocks based on 2:2:1 configuration. This will involve the addition of four Heat Recovery Steam Generators (HRSG) and two steam turbines to the power plant. The capacity of the power plant will increase from 440 MW to 660 MW. It is also envisaged that after conversion to the combined cycle operation mode the status of the power plant would change from peaking plant to the base load plant depending on the operational requirements of the overall generation system.

FUTURE PERSPECTIVE OF IPP'S IN MALAYSIA

The success in the privatisation of TNB and the role played by IPP's in alleviating the power shortage in the country has prompted the Government of Malaysia to continue the grant of licences to the independent power producers consistent with the general objectives of the privatization. Announcements have been made to issue licence to YTL Power Generation to construct power plant in Johor state to supply power to Singapore and to Perlis Power for construction of a gas-fired power plant to serve the energy needs of Southern Thailand. In Sabah, IPP projects which are under construction include two power plants with capacities of 50 MW and 120 MW.

With the rapid development of the privatization of power industry there is an urgent need to harmonize or standardize the many forms of power purchase agreements currently applicable. It is also imperative that agreements should improve on provisions based on realistic assessment of risks, securities and warranties to be borne or provided by future IPPs.

Taking into account the differences in the characteristics and capital investments applicable to each IPP, it is desirable to finalize the ground rules for removing the existing disparities in power tariffs for all IPPs.

REFERENCES

1. Muzaffar Tate, 1989, "Power Builds the Nation, Volume 1 - The Formative Years" , National Electricity Board.

2. Muzaffar Tate, 1990, "Power Builds the Nation, Volume 2 - Transition and Fulfillment" , National Electricity Board.

3. Lim Ewe Jin, 1994, " Private Power Generation in Malaysia: An Independent Power Producer's Pespective" , Second Afro - Asian Conference on Power Generation, Transmission and Distribution , Kuala Lumpur.

4. Association of IPP's, 1995, "Power Talk", A triannual publication of Association of IPP's, Kuala Lumpur, Malaysia.

Acknowledgement

The authors wish to gratefully acknowledge the generous assistance provided by Ir. Lt. Cdr. (B) Hj. Mohd. Ismail Hj. Che Mat Din, Managing Ditector of Port Dickson Power Company, in the preparation of this paper.

PRIVATE POWER GENERATION FROM THE PERSPECTIVE OF A DEVELOPING COUNTRY: A CASE STUDY OF HONDURAS

J. Williams, F. Mayaki

SNC-LAVALIN Inc., Canada

INTRODUCTION

Background

The trend throughout the world is towards restructuring, deregulation and increased involvement of the private sector in the generation and distribution and, to a lesser extent, in the transmission of electricity. The trend can be observed in OECD countries such as Britain, Norway and New Zealand, but also in the less developed countries such as Chile, Argentina, the Philippines and many others

This conference is aimed specifically at electric power generation and it is on the generation aspect of the power industry that this paper focuses. The viewpoint however is that of the developing country rather than the independent power producer. In taking this approach it is believed that useful insights may be provided as to what strategies may be valid for independent power producers as they approach the power generating markets of developing countries. Rather than discuss developing countries in general, the paper considers the specific case of Honduras. While each country must be considered as an unique case, a number of the issues being faced by Honduras, many of which are yet to be resolved, can be expected to occur elsewhere.

Forces that Drive the Reform Process

At times it seems that the drive towards restructuring and privatization is so widespread as to be the one identifiable trend of the 1990s in the electric power sector. Many countries have considered the restructuring of their electricity systems, opening up generation to the independent power producers and offering open access to the transmission system for the new generators and larger consumers. Treating the phenomenon as a fashionable trend however is not sufficient; consideration needs to be given to the driving forces that are promoting reform in the developing countries.

The influence of the OECD countries is significant. The fact that a country like Britain can break up the Central Electricity Generating Board, relatively successfully, carries considerable weight. The idea is promoted that there are areas which are better handled by the private sector than the public sector and that publicly owned and operated utilities are generally inefficient. There is increasing acceptance that only parts of the integrated electricity industry are a natural monopoly, notably the transportation aspects of transmission and distribution.

Such trends in thinking are passed on to the less developed countries (LDC) through the education their future professionals receive in OECD countries and, through the acceptance and promotion of these ideas, by the international funding institutions such as the World Bank and other regional development banks.

The influence of the International Funding Institutions (IFI) is considerable. Throughout the 1960s, 70s and into the 1980s, these agencies supported the development of national, publicly owned, integrated electric utilities through the funding of projects. Often projects could only be built with the support of the IFIs, whose endorsement attracted any additional private commercial capital required to complete projects. Since the late 1980s however, there has been considerably less interest in the funding of power projects in the public sector on the part of the IFIs. In addition to withdrawing the availability of funding from generation projects, pressure is exerted by making compliance with IFI policies in the energy sector a condition of disbursement of funds in other sectors such as economic reform and modernization-of-the-state projects.

Shortage of capital is a major force that promotes reform, in part due to the withdrawal of IFI project funding support, but also due to the historically poor financial performance of the electric utilities. Poor financial performance makes it difficult for the electric utility to raise capital and further forces the developing countries towards opening up power generation to the private sector.

Opportunities and Advances in International Power Generation, 18–20th March 1996,
Conference Publication No. 419, © IEE, 1996

This poor performance is often a mixture of inadequate tariffs, weak management and high debt. Political interference in the setting of tariffs is common and independent regulation of tariffs is rare. Tariffs are often kept low for a mixture of social and political reasons. Many developing utilities have high debt due to the investments that were made at the time they were supported by the IFIs. Such investments often involved the building of large hydro-electric projects based on studies which, at the time, showed them to be the least economic cost options.

High debt, usually denominated in foreign currencies, requires debt service to be covered by earnings from electricity sales which are entirely in local currencies in the absence of export sales. Failure to adjust tariffs in line with the prevailing exchange rate leads to deteriorating debt service ratios.

Shortage of funds affects the utilities of developing countries in ways other than in their ability to raise capital for future projects. Lack of funds, particularly foreign exchange, leads to inadequate maintenance due to lack of spare parts. A combination of a lack of funds and public policy regarding salary levels means that salaries are low in comparison to the private sector, resulting in low employee morale and loss of trained technical staff to the private sector.

Inefficiencies, both perceived and real, of national vertically integrated utilities, also lead to pressure to reform. The reform is often led by the IFI but eventually receives support from the national government due to the adoption of the IFI view by local bureaucrats and the pressure exerted by the IFIs.

In the recent past, electric utilities in developing countries tended to be publicly owned and operated. They were often run at a loss due to a combination of inadequate tariffs and poor management.

The systems were frequently characterized by high technical and non-technical losses (including theft, poor metering and inaccurate billing). The low energy consumption of the average consumer combined with his limited ability to pay put further pressure on the tariff level and on profitable operation.

Direction of Reform

The general direction of reform usually results in the breakup of the vertical integration of the national utility, with generation, transmission and distribution separated from each other.

Corporatization of each sector is promoted as well as privatization options, particularly of the generation function. Competition is considered desirable but may not be easily achieved due to the lack of competent players required to operate in an environment perceived as risky. A true market system in the British or Norwegian sense is not easily achievable and unlikely to evolve for many years. Competition may only be possible at the stage that contracts to supply electricity and operate plant are being awarded.

Transmission is generally accepted as being a natural monopoly while some attempt is made to regionalize, corporatize and sometimes privatize the distribution system. More open access to the transmission system is encouraged as well as to the "wires" aspect of the distribution system so as to unbundle the services being provided by the distribution companies.

Implications for the Less Developed Countries

Reform of the electrical sector means that the existing national utility is broken up, at least to some extent. The old rules are replaced and the national utility is subjected to the pressures of competition. The area most likely to be affected in the early stages of reform is the supply of new generation. Existing generating plant may also be offered for sale, or contracts may be awarded for rehabilitation and operation of existing plant.

Increasingly, there is pressure to distance government from the utility and it operations. Tariffs in particular become less tied to the political process and an attempt is made to provide independent regulation. In countries which have no tradition of such separation the transition can be expected to be difficult. The temptation to influence tariffs for social reasons or for political gain may remain strong.

Regulation itself is an entirely new area for most developing countries. Little local experience can be drawn on and the presence and influence of the IFIs is strongly felt. In many developing countries the IFIs are prominent players in the development of reform legislation and subsequent regulation methods and procedures.

Ownership of future generation plants may be expected to be in private hands. Distribution companies may end up in private hands or in companies of mixed public and private ownership, with local cooperatives and municipalities becoming players in the electricity sector.

THE CASE OF HONDURAS

Honduras

Honduras, a Central American country classified by the World Bank as low income, has a population of 5.3 million and a per capita GDP of US$600 in 1994. Its main exports are bananas and coffee with a recent emergence of shrimp farming.

Empresa Nacional de Energía Eléctrica (ENEE)

Until 1994, ENEE was solely responsible for the generation, transmission and distribution of electricity in Honduras. The maximum demand in 1994 was 482 MW and some 442 000 consumers were served. Per capita consumption of electricity is low at 515 kWh/year compared with 5700 kWh/year in Britain and 18 300 kWh/year in Canada. Losses in the system are high at 27% of net generation and are split almost evenly between technical and non-technical.

Until 1994 the system was predominantly hydro based with 515 MW installed capacity and some 95% of energy generated by hydro plants. One plant, El Cajón, commissioned in 1985 represented 300 MW or 58% of the total installed capacity of the country. Between 1985 and 1993 Honduras was a net exporter of electricity.

Over 40% of the national debt of Honduras is due to the electric power sector with the El Cajón project accounting for most of this debt.

The 1994 Electricity Crisis

Honduras suffered a severe crisis in its electricity supply industry in 1994. Power cuts of over 12 hours per day had to be implemented for several months. Local estimates have placed the losses to the economy at up to 10% of GDP for 1994. While this figure is probably overstated, it indicates the magnitude of the problem and the impact it had on the Honduran economy.

What brought the problem about was a combination of a dry year, inadequate installed capacity and management of the reservoirs. The over-installation of the 1980s had placed severe pressure on ENEE in terms of debt service. As with all LDC utilities, debt was mainly denominated in hard currencies while practically all receipts were in local currency. The El Cajón project had been

partly justified by projected sales of surplus power during its early years in service. However, attempts to export surplus energy were unsuccessful as much of the energy exported was never paid for. National tariffs failed to keep pace with ENEE's financial needs with the results described above.

Political pressure to keep the lights on in spite of the low reservoir levels compounded the problem as the energy value of stored water decreased with decreasing reservoir levels.

Lack of funds in the years leading to the crisis produced a number of problems. Maintenance of existing diesel plant was neglected, so that when these plants were required for generation they were not available. Lack of maintenance and poor management of the distribution system combined with low ability to pay led to losses of around 27% of net generation. Low salary levels led to low morale and to the loss of staff.

Under these panic conditions Honduras went to the market to purchase power from the independent power producers (IPP). The initial attempts were unsuccessful due to the limited response received from the producers, in part due to inadequate tender documents but also due to the risk aversion of the private developers. Eventually three contracts were signed for new generating plants, one of which subsequently failed.

New Generating Plants

Project X involves the purchase of 80 MW of diesel generation installed in three stages, 24 MW initially, 60 MW and eventually 80 MW. The contract was negotiated by ENEE, who at the time had little previous experience of negotiating power purchase contracts. The resulting contract resulted in an energy purchase price of 8.5 US cents/kWh. The contract was not specifically *take-or-pay* but the purchase price included fixed payments which resulted in a de-facto *take-or-pay* contract at a capacity factor of 55%. The independent power producer is responsible for the supply of fuel and the contract includes a fuel price adjustment clause based on international prices. The fuel is actually purchased in the local market where fuel prices are protected. The projected average energy purchase price for 1995 is US cents 11.0/kWh.

Project Y is a *rehabilitate, operate and maintain (ROM)* contract. This involved the rehabilitation of 80 MW of existing ENEE diesel sets that had fallen into disrepair due to lack of maintenance. Fuel for these sets is provided by ENEE. Payment for energy is based on a combined fixed and variable fee that effectively ensures that the units must run. In effect it is a *take-or-pay* contract.

Project Z was a contract for the purchase of 70 MW of power generated by gas turbines. Even though the contract was signed and gazetted in the government press, the project did not go ahead due to the failure on the part of the contractor to provide the necessary performance bonds and guarantees.

THE RISKS TO HONDURAS

Lack of Competition

The lack of competition manifests itself in many ways. In the early stages of opening up generation to the private sector, there is a shortage of investors willing to accept the political and financial risk. The lack of precedence contributes to the uncertainty, including how contracts will be interpreted and questions of repatriation of profits. Under these conditions, IPPs prefer sole source, long term contracts with guarantees, while the LDC is seeking multiple offers to check that the prices being offered are competitive.

Competition of the type seen in Britain and Norway is not a realistic option for developing a relatively small electrical system in the near term due to the small size of the market, the small number of players and the fact that the infrastructure and sophistication required for market trading are not available. Competition may therefore only occur at the time that contracts are initially awarded, which excludes sole sourcing.

Lack of Expertise and Weak Regulation

Negotiation with a sole source contractor can lead to bad contracts not only due to lack of competition but also due to the inexperience of the negotiating team from the LDC's utility.

At the onset of restructuring, the LDC utility and associated government departments involved in the energy sector have little experience of negotiating power purchase agreements. The concept of regulation may be new with little national expertise available. The utility operates under a new set of rules in an environment in which it has not previously participated. Inexperience may lead to poor decisions.

The independence of the regulatory process may be difficult to establish in a regime where the utility has not operated autonomously and has been strongly influenced by political pressures and

decisions. Considering that politicians may not easily give up powers that they once had, this leads to an uncertain operating environment and to the continued risk aversion of the IPP. The area most affected is the setting of tariffs which is often a hot social and political issue, particulary as elections loom.

The Risk Aversion of the IPP

Under conditions of uncertainty the IPP will look for projects with minimum capital exposure and rapid cost recovery leading to least capital cost options with short construction periods which essentially become thermal plant, usually medium speed diesels. The IPP will also seek long term contracts with guarantees.

Weak Bargaining Position of LDC

The LDC is in a weak bargaining position. When negotiating with contractors Honduras was desperately short of power. The need for power was immediate with the IPP looking for long term, take-or-pay, guaranteed rates of supply and protection against future risk of fuel price increase and exchange rate variation.

Lack of Freedom to follow the Least Cost Plan

Honduras will continue to plan its generation at the national level to produce the so-called "indicative plan". Increasingly however it is the private sector that will decide what will be built.

As stated earlier, unguided, the IPP will be biased towards low cost, fast payback projects, that is to say thermal plant. Complete protection against fuel price increases reinforces this bias. Honduras therefore has to devise methods that allow other types of plant to be built if this is what the indicative plan recommends.

The history of fuel prices since 1973 demonstrates clearly that the doubling and tripling of fuel prices has occurred. There is no reason to believe that this cannot be repeated.

Honduras may have to implement a process where the free market is guided as to what plants will be built. A partial answer may come from reducing the level of protection that the IPP receives regarding future fuel prices. This would force the IPP to assume some of the responsibility for fuel price risk and would ensure, at least to some extent, that the IPP takes future fuel prices into account in the investment decision, allowing other plants such as hydro, geothermal, bagasse and wind to be considered.

Another approach may be for Honduras to indicate to the market what type of plants it is willing to consider. Typical projects might be a 25 MW wind farm, a 30 MW geothermal development or a 55 MW hydro plant at a particular location.

This approach implies a degree of national planning. It also implies that random offers to produce power from individual private suppliers would not be considered. Rather, bids would be sought for the development of a particular resource or site. Invitations to tender for such projects would be offered to prequalified bidders. Developers would be forced to bid against each other in a competitive situation, with the award of contract basically decided on their selling price for the electricity produced.

Liberalization and Restructuring of the Market

In addition to opening up generation to the private sector, Honduras is allowing freer access to the transmission network and, to a lesser extent, the distribution system. This brings advantages to the consumer, particularly the larger consumers who are free to purchase their electricity from different sources. It also brings a certain degree of risk. Free access to the transmission may also allow the IPP developer to cream off the larger profitable consumers leaving the loss making consumers to be served by ENEE. Tariffs to the loss making consumers would inevitably rise.

Power purchase contracts that cover the IPP's fixed costs can also be potentially problematic. Once the fixed costs are covered the IPP may be able to sell profitably at a level slightly above his variable costs. The lower cost "surplus" power could be sold to large consumers or even internationally.

Leaving large consumers connected to the ENEE system, which will normally be the case, also leads to potential problems. The large consumer may rely on ENEE to provide the reserve for the system, buying only energy from the IPP. This leaves ENEE with costs but no compensating revenues.

The Privatization Trap

The privatization trap arises when the private sector distrusts the regulatory process. Because of mistrust, the IPP seeks a higher risk premium in his tariff. The resulting high tariff allows the IPP to earn returns which the regulator eventually regards as above average. The regulator in turn comes under pressure, both public and political, to amend the tariff downwards. Acting to reduce the tariff confirms the IPP's perception of an unfair regulator.

Certainly the contracts signed early on in Honduras have resulted in high tariffs. Pressure to renegotiate these early contracts can be expected. Opening them up however leads future investors to demand an increase in the risk premium to compensate for this additional perceived risk and thus the cycle continues.

The temptation for the IPP is to negotiate the best deal available at the time. With no competition and in a situation where the LDC is short of power, sales at above market rates may be made. Eventually however, the regulator will come under strong pressure to intervene in the process. In the long term, an unfair contract is in the interest of neither the IPP nor the LDC.

FUTURE DIRECTIONS

Evolution of the Honduran Electricity Market

There can be little doubt that IPPs will increasingly participate in power generation in Honduras. The initial plants will be thermal but may eventually include other types of generation, including hydro, geothermal, bagasse and wind, particularly if the direction is influenced by the regulator.

At the same time increased access to the transmission system will be allowed which will permit direct sales from the IPP to large consumers using the national grid to transmit power.

Increasingly it can be expected that Honduras will attempt to influence what is built. This might be achieved by seeking bids for specific sources and sites. This implies that Honduras will have to invest in feasibility studies, particularly for hydro sites. There will be a move away from sole sourcing and from granting exclusive site rights. Reduction in the degree of protection for fuel price increase might also be expected.

The competence of the regulatory regime can be expected to increase with time and, as the regulatory body gains the confidence of the industry, will reduce government interference. If this does not happen, Honduras can expect to have to continue to pay a risk premium for the uncertainty that the involvement of government in the setting of tariffs implies.

Strategies for Power Project Developers

This paper has discussed the issue of private power generation in Honduras from the perspective of the host developing country. It is believed that this approach provides insights that may help private investors to better understand, plan, design and negotiate their projects and contracts. In designing strategies for power projects in developing countries, the following factors should be considered.

Identify projects that are in the national interest. Even if the process of developing projects and negotiation may take considerable time, the possibility of being the successful developer of a project of national interest may well offset and compensate for other commercial and operational risks. This strategy implies dealing with national long term priorities rather than risk a situation that sooner or later becomes a source of conflict between the IPP and LDC such as when high fuel price changes occur. Avoiding potentially conflicting situations with the regulator or the government who under political or social pressures are obliged to intervene, may reduce considerably the business risk for the IPP.

Broaden the potential market for the power generated as a means for ensuring the long term viability and reducing the risk of the project. Rather than solely looking for sovereign guarantees of the Honduran state, identify buyers other than ENEE. These will include large consumers, future decentralized distribution companies and exports to the de-integrated markets of neighbouring countries. The existing interconnected grid will help to secure the financial viability of the project as well as spread the risk across the regional market. Thorough knowledge regarding the reforms that are taking place across the region is required to take advantage of these risk reducing possibilities. The diversity of the load shapes in the region will also tend to improve the flexibility to operate within this new regional market.

Include local inputs and knowledge in the projects. Local partnerships and hiring local highly skilled staff and business managers will certainly help. Local hiring of technicians and labour will also assist in gaining acceptance in the community. Local know-how is invaluable in resolving problems. Such means will be particularly efficient and acceptable if they are perceived as contributing to the economy and welfare of the community, country and region. Developing good public relations and marketing the project locally are important strategies in gaining local acceptance of the project. This is particularly important wherever

relocation of people may be involved and where there may be an environmental impact. Maximizing local investment content and making use of local funding will help reduce future foreign exchange risks.

Be Flexible and Innovative. Flexibility on contractual conditions in the area of economic dispatch of plant may well provide greater access to other markets by allowing surplus energy to find markets. Rigid application of contract conditions may prevent the sale of this surplus, marginal cost, energy. Innovation in the pricing of energy to take into account different values depending on time of day and season may also lead to new market opportunities.

Ensure that contracts appear fair with respect to what is taking place in the neighbouring countries and accommodate for flexibility. For similar risks, consumers will expect similar energy prices. As the Central American countries move towards regional markets, the pressure to renegotiate contracts which appear unfair will increase. Flexibility in the discussion of existing contracts which provide windfall profits may reduce current energy sales prices but may well be good long term investments in building long term relationships, confidence and trust.

key reference
Plan Maestro de Energía Eléctrica, Empresa Nacional de Energía Eléctrica, Honduras, Sept 1994.

GLOBAL CLIMATE CHANGE AND HYDROELECTRIC POWER

H. W. Whittington, S Gundry
University of Edinburgh, Scotland

Abstract

As part of overall Global Climate Change (GCC), changes in levels and distribution of precipitation will affect the viability of existing and proposed hydro electric stations.

GCC models predict changes in precipitation in many regions, both in terms of annual mean levels and seasonal variability. Such changes are forecast to be significant in the year 2050 and beyond. In areas where precipitation is reduced, existing hydroelectric installations may require early replacement with alternative sources of power . As economic appraisals of new hydro electric projects are normally based on a 60 year (or greater) life, GCC will also necessitate revisions in assessments of potential hydro electric capacity.

Introduction

Power from running water has been exploited for several thousand years by man and over the past hundred or so years has been available in the form of hydro electricity. The technology is mature and the resource comes in the category of so-called renewables. As such, hydro electricity has many attractive features: these include virtually zero gaseous pollution, no imported fuel charges and a long life for the plant, especially the civil works. The main adverse effects of large scale projects are the impacts upon local agro-ecosystems, both in the catchment itself and downstream of the reservoir.

Energy demand in developing countries, including those in the former CIS and Eastern Europe, is expected to increase sharply over the coming 20 years. If these increases are met from new fossil fuel plants, the output of carbon dioxide to the atmosphere will exacerbate GCC. As an alternative, many of these countries possess river catchments which, on the basis of current weather patterns, have high potential for production of hydro electricity and offer renewable and non-oil dependent fuel source of power generation.

The conditions for successful exploitation of hydro electric power exist in many parts of the world. The exploitation of a steady and reliable resource of running water can be improved by the provision of pondage (a reservoir). However, the feasibility of any scheme depends on the hydrology of the region which has two main components:

- Catchment area of the planned hydro electric station;
- Precipitation levels within that catchment area.

The catchment area of any planned station is determined by the topography and locations of existing water sources, both above and below ground. GCC will not alter the locations of these within the planning horizon of most new projects. However, the hydrological analyses of catchments, produced from historical records of precipitation, will need to be revised. These analyses, which are available for most countries, form the bases of the estimates of potential hydro electric capacity, as used the World Bank and others.

Climate Change

The earth's climate is a complex system, affected both by natural events, such as volcanoes, and by man-made interference. Increases in the atmospheric concentration of carbon dioxide (CO_2), bring about an increased heating of the Earth through radiative forcing - 'the greenhouse effect'. Since the late eighteenth century, levels of CO_2 and other greenhouse gases have increased due to man's activities producing widespread release of gaseous emissions into the atmosphere. Major sources of such emission include transport networks and the burning of fossil fuels to generate electricity. Recent reports suggest that increases in these emissions since the pre-industrial era make the largest individual contribution to radiative forcing (Houghton, 1995). Small release of some complex chemicals, such as chlorofluorocarbons (CFC's), have an exacerbating impact upon the changes by altering the chemistry of the upper levels of the atmosphere. Furthermore, the destruction of large areas of plant biomass e.g. rain forests, has reduced the Earth's capacity to absorb CO_2 from the atmosphere.

Opportunities and Advances in International Power Generation, 18–20th March 1996,
Conference Publication No. 419, © IEE, 1996

The result of these factors is that warming of the Earth's climate is set to continue into the next century, regardless of current policy, until anthropogenic greenhouse forcing reaches a new equilibrium value. The magnitude of this new equilibrium value is the subject of much research and will depend upon, inter alia, the increases in atmospheric water vapour (Raval and Ramanathan, 1989), reductions in surface snow and sea-ice cover (Ingram et al, 1989) and cloud distributions (Wetherald and Manabe, 1988; Roeckner et al, 1987; Senior and Mitchell, 1992).

In addition to determination of the new equilibrium value, which is time-independent, researchers are investigating the likely transient response of the climate towards this new equilibrium (Murphy, 1992). Of particular importance is the role of the oceans in providing thermal inertia to the overall system (Hoffert et al, 1980). Clearly if atmospheric warming could be transferred through some sort of heat sink into the deep oceans, the period to reach the new equilibrium would be many hundreds of years. Conversely, if the warming is limited to the upper levels of the oceans, the equilibrium will be reached quickly. In order to model this transient response, researchers have developed models which couple the atmospheric changes to the oceanic flows. Such models are known as 'Atmosphere/Ocean Global Coupled Models' (AOGCM), of which the Hadley Climate Change model, is the main UK experiment.

The Hadley Climate Change Model

The Meteorological Office has produced a report (Murphy, 1992) on global climate change. The assumptions made within this study include an allowance for a compound increase of carbon dioxide concentration of 1% per annum. The study looks forward some 70 years and assesses the potential effect of global warming due to the greenhouse effect. Of particular interest in the present context are the predicted changes in precipitation characteristics around the world. Figure 1, taken from the Hadley Report, shows, in coarse summary form, the expected changes in precipitation in 50 years time. It should be noted that the major changes are predicted to occur in the Northern Hemisphere, since it is here that land mass predominates (in the Southern Hemisphere, temperature changes are minimised by the large masses of ocean which circulate both on a surface level and to considerable depth absorbing the increase in heat available). Within the Northern Hemisphere, the changes in precipitation are maximum in the centres of

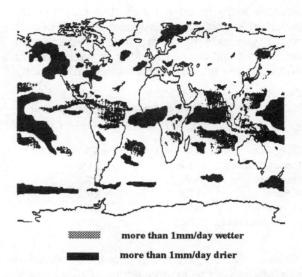

more than 1mm/day wetter
more than 1mm/day drier

Figure 1. Mean annual precipitation changes over the next 50 years (Ref: Hadley Centre)

large land masses, well away from the moderating influences of the oceans. For example, the changes predicted for the United Kingdom are relatively small and in themselves, should not create particular difficulties for operators of hydro electric plant. However, in two regions which form the basis of the case studies of this paper, the effects predicted are much greater and could lead to significant reductions in water run-off for particular hydro electric schemes.

It should be noted that other AOGCM take different views of precipitation impacts of GCC. Hadley suggests that the same quantity of rainfall will be re-distributed, whereas other models suggest that re-distribution will be accompanied by an overall increase in precipitation due to the increased moisture levels in the atmosphere.

Case Study: Central and Eastern Europe

Within the former Soviet Union hydro electric power is well developed: for example, the rivers of the Volga and Dneipir have been extensively exploited (Whittington, 1990). However, it is a relatively underused asset in the Satellite States of the Eastern Bloc,.

Table 1 gives details of estimated hydroelectric in the countries of the former Soviet Bloc. Exploitation of reserves to date has been limited to less than one third. Of particular interest is the former Yugoslavia, where the rebuilding of national infrastructure after the hostilities could involve substantial expansion of the largely unexploited hydroelectric resource of the region..

Figure 1 suggests changes of around 1mm/day in average precipitation in central and eastern

Europe. More detailed maps (too complex for inclusion in this paper) suggest that all of this region will experience significant reduction in annual precipitation (1mm/day is equivalent to 14 inches per annum). To a first approximation,

Country	Potential Hydropower (TWh)	Potential Hydropower already used (%)
Bulgaria	3.0	NA
Czech/Slovak	10.8	31%
Hungary	6.0	2%
Poland	12.0	26%
Romania	40.0	33%
ex Yugoslavia	170.9	16%

Table 1. Eastern European countries - installed electricity generation and potential hydropower resources (Ref: CEC DGXVII, 1994)

for the same catchments the change in hydrology may be taken as merely the ratio of main precipitation levels. The authors realise this approach may be somewhat simplistic in that the changes in precipitation may not be uniformly distributed and that the effects may be significantly different from region to region. However, if such a comparison is made, with a typical level of 700 mm per annum, it can be seen that reduction of around 50% in hydro resource might be expected.

Case Study: Africa

Although the hydro electric schemes of the Nile (the Aswan dam in Egypt) and Zambezi (the Kariba dam , Zimbabwe and Caborra Bassa, Mozambique) are perhaps the most well known projects, the catchments of the Zaire river and its tributaries have greater potential for hydro electric generation than the rest of Africa combined. In 1993, the African Development Bank funded a feasibility study to build a 4,000 kilometre power line to transmit electricity from Zaire to Egypt via the Central African republic, Chad and the Sudan. The South African government has considered a similar scheme to draw power from Zaire to the industrial and growing domestic fuel consumers in Southern Africa. It can be seen from Figure 1 that the Zaire basin is expected to suffer a reduced precipitation level of approximately 400 mm per year or about 20% of the current/historic levels. Such a change would produce a significant impact upon the economic viability of hydro electric schemes and the associated long distance transmission lines.

Table 2 below shows the hydro electric capacity for countries in the Southern African Development Community (SADC). As can be seen, these countries possess substantial

potential capacity considerably in excess of their current installed capacity.

Country	Installed capacity (MW)	Potential Hydropower (MW)
Angola	466	17,680
Botswana	212	Not known
Lesotho	3	360
Malawi	144	660
Mozambique		
- domestic	150	12,840
- exports to SA	2,195	
Namibia	360	2,700
Swaziland	51	120
Tanzania	440	5,590
Zambia	1,750	5,810
Zimbabwe	1,961	2,920

Table 2. SADC countries - installed electricity generation and potential hydropower resources (Ref: Junior, FVL, MSc thesis, University of Dundee 1992)

As these countries expand their economies and also export electricity to support South Africa's expansion, their medium term plans for electricity generation are necessarily centred on hydro resources. Any material change in these resources resulting from GCC would require a revision of such plans.

Economic Considerations

Winje (1991) has noted the important role of electric power in the economic growth and improvement in living standards of developing countries. Such countries, excluding China, account for 16 % of global primary energy consumption, of which about 30 % is used for electricity generation. This is expected to increase to 50 % in the next century. This expected growth in electricity generation will confront the developing nations with technological, environmental and economic problems. It is suggested that the economic problems will be significant worsened if expansion of hydro generation proceeds without recognition of the impact of GCC upon current estimates of potential capacity.

Generating costs for any hydro electric scheme are primarily a function of the capital cost of the civil works, the power generation and transmission systems. For locations remote from the eventual market for the electricity, the transmission costs can often be the largest element. Capital costs are amortised or depreciated over the assumed life of the scheme, which is generally 80 years for the civil works and 35-40 years for the electromechanical equipment. An indicative figure would be £1,000 per kW installed. As fuel costs are zero and running costs minimal, operation and maintenance is approximately one penny per kW hour. Where stations are operated beyond this assumed 'book' life, the capital element of the generating costs is removed and the cost per kW hour falls sharply.

The economics of plant installation in the Soviet Bloc over the past 70 years are more complex than those in the West, in that many projects went ahead justified on strategic or political grounds rather than straight forward economic cost grounds (Whittington, 1990). However, again, for illustrative purposes, it will be assumed that traditional economic analysis used in Western market economies will be applicable to installations in this part of the world. This is partly true for present installed plant and, with the break-up of the Soviet system should be true for all new plant being installed.

For the developing countries, particularly in Africa, additional cost considerations may apply related to the source of funds used for the initial capital expenditure. Capital costs of civil and electro-mechanical plant are substantial and can only be funded by large loans from global financial institutions such as the World Bank. These loans are usually denominated in hard currencies and interest and capital repayments present a major, long term foreign exchange exposure to the fragile economies concerned. With the adverse movements in exchange rates suffered by developing countries in the last 30 years, the result can often be a debt burden for hydro electric schemes which cannot be even serviced from the sales of generated electricity, let alone begin to repay the capital borrowed.

Conclusion

Using renewable resources to generate electricity makes hydro electric schemes inherently attractive to the expanding economies of the Soviet Bloc and other developing countries. However, the economics of such schemes are wholly dependent upon the long term appraisal of capacity to support capital investment. In addition to this high front-end loading, long term commitments to high levels of debt servicing in scarce foreign exchange are usually required.

It is the authors' view that GCC indicates that the potential hydro electric capacity of these countries requires review and that the implications of any changes in such capacity upon their predicted economic development should be analysed.

Bibliography

Commission of the European Communities: 'The European Renewable Energy Study' DGXVII, Main Report, ISBN 92-826-6950-5, 1994.

Hoffert M.I., Calligari A.J. and Hsieh C-T., 1980. The role of deep sea heat storage in the secular response to climatic forcing. J. Geophys. Res., 85, 6667-6679.

Houghton J.T., Meira Filho L.G., Bruce J., Hoesung Lee, Callander B.A., Haites E., Harris N. and Maskell K., 1995. Climate Change 1994: radiative forcing of climate change and an evaluation of the IPCC IS92 emission scenarios. Cambridge University Press on behalf of the Intergovernmental Panel on Climate Change.

Ingram W.J., Wilson C.A. and Mitchell J.F.B., 1989. Modelling climate change: an assessment of sea-ice and surface albedo feed backs. J. Geophys. Res., 94, 8609-8622.

Murphy J.M., 1992. A prediction of the transient response of climate. Climate Research Technical Note 32, Hadley Centre, Meteorological Office.

Raval A. and Ramanathan V., 1989. Observational determination of the greenhouse effect. Nature, 342, 758-761.

Roeckner E., Schlese U., Biercamp J. and Loewe P., 1987. Cloud optical depth feedbacks and climate modelling. Nature, 329, 138-140.

Senior C.A. and Mitchell J.F.B., 1992. CO_2 and climate: the impact of cloud parameterization. J. Climate.

Wetherald R.T. and Manabe S., 1988. Cloud feed back processes in a general circulation model. J. Atmos. Sci., 45, 1397-1415.

Whittington H.W., 1990, 'Hydroelectric development in USSR' IEE Proc Part C, pp 343-348.

Winje D., 1991. Electric power for developing nations. In 'Energy and the environment in the 21st century', Tester J.W., Wood D.O. and Ferrari N.A. (Eds.) 611-620. Procs. of Conf. at M.I.T. March 26-28, 1990. MIT Press.

Junior, FVL. 'Matrix Management for SADC Electricity Projects'. MSc U of Dundee, 1992

POWER LIMITATION IN VARIABLE SPEED WIND TURBINES WITH FIXED PITCH ANGLE

R. Cárdenas   G.M. Asher   W. F. Ray   R. Pena

Electrical & Electronic Engineering Department. University of Nottingham. Nottingham. England.

1 INTRODUCTION

Variable speed operation of wind turbines has the potential to increase energy capture, to reduce fatigue damage and to reduce aerodynamic noise level. However one of the most important characteristic of variable speed wind turbines is operational flexibility, which allows the use of different control strategies for above and below rated wind speed. One of the control strategies which is possible to implement is power capture limitation without using pitch control.

Figure 1 shows the power characteristic of a wind turbine in the stall region, it can be seen that when the wind speed increases from V_1 to V_2, the wind turbine has to be driven to the point P_2 in order to keep the power in the nominal value. When the wind speed increases even more to V_3, the wind turbine has to be moved again to the point P_1 in order to keep a constant power.

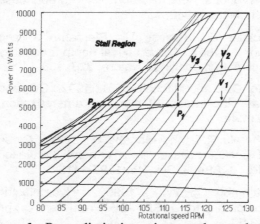

Figure 1.- Power limitation using speed control.

From Figure 1 it is seen that the power limitation can be achieve using a speed control system which change the speed of the machine in order to change the captured power.

This alternative for power control has been suggested in Goodfellow and Smith [1], Connor and Leithead [2] and implemented with a 20Kw wind turbine in Thiringer and Linders [3]. The main advantage of this method is that the machine can be operated in all the wind speed range with a fixed pitch blade.

In this paper, a control system for power limitation

driving the wind turbine to the aerodynamic area is proposed and experimental result obtained from an experimental rig which uses a high performance switched reluctance generator are discussed.

2 POWER VARIATION WITH ROTATIONAL SPEED.

According to Figure 1, there is a variation of power which is produced by a variation in the rotational speed. Mathematically the power captured by a wind turbine can be written as:

$$P_m = \frac{1}{2}\pi\rho C_p(\lambda)R^2V^3 \qquad (1)$$

Where λ is defined as:

$$\lambda = \frac{\omega R}{V} \qquad (2)$$

Where ω is the rotational speed, V the wind velocity, R the wind turbine radius and C_p the power coefficient which is a function of λ. The small signal equation for (1) is:

$$\Delta P_m = \frac{\partial P_m}{\partial \omega}\Delta\omega + \frac{\partial P_m}{\partial V}\Delta V \qquad (3)$$

Using the chain rule, the variation of the power respect to the rotational speed is:

$$\frac{\partial P_m}{\partial \omega} = \frac{\partial P_m}{\partial \lambda}\frac{\partial \lambda}{\partial \omega} \qquad (4)$$

Using (2) and (4):

$$\frac{\partial P_m}{\partial \omega} = 0.5\rho\pi R^2V^3\frac{\partial(\omega R/V)}{\partial \omega}\frac{\partial C_p}{\partial \lambda} = \frac{1}{2}\rho\pi R^3V^2\frac{\partial C_p}{\partial \lambda} \qquad (5)$$

The aerodynamic torque produced by the wind on the blades is:

$$T_m = \frac{1}{2}\rho\pi C_t(\lambda)R^3V^2 \qquad (6)$$

Where $C_t(\lambda)$ is called the torque coefficient.The relationship between the torque and the power coefficient is:

$$C_t = \frac{C_p}{\lambda} \qquad (7)$$

Using (6), (5) can be written as:

$$\frac{\partial P_m}{\partial \omega} = \frac{T_m}{C_t(\lambda)}\frac{\partial C_p}{\partial \lambda} \qquad (8)$$

Using (7) in (8), the variation of the power respect to the rotational speed is:

$$\frac{\partial P_m}{\partial \omega} = T_{mp}\left[\lambda\frac{\partial \ln(C_p)}{\partial \lambda}\right] \quad K_\omega = T_{mp}\left[\lambda\frac{\partial \ln C_p}{\partial \lambda}\right] \qquad (9)$$

and the small signal variation of the power with the

Opportunities and Advances in International Power Generation, 18–20th March 1996, Conference Publication No. 419, © IEE, 1996

---

rotational speed is obtained by replacing (9) in (3). The term T_{mp} corresponds to the value of the mechanical torque at the point where the system has been linearized. If it is assumed that the objective for above rated wind speed is to obtain a constant power equal to the nominal value of the generator, then the value of (9) for each wind speed can be obtained from the operation curves shown in Figure 1.

According to (3) there is also a variation in the captured power due to a change in the wind speed. This power variation can be calculated as:

$$K_v = \frac{\partial P_m}{\partial V} = K_p \frac{\partial (C_p(\lambda)V^3)}{\partial V} = K_p V^3 \frac{\partial C_p(\lambda)}{\partial V} + 3 K_p C_p(\lambda) V^2$$

$$(10)$$

Where K_p are the constant term of (1).

The terms inside bracket in (9) and part of K_v defined in (10) are dependant on the characteristic of the blade being used. In general blades with a narrow and sharp peak C_p characteristics have a large variation of power with small variation on the rotational speed. Figure 2 shows two C_p-λ characteristics, one of them with a narrow and sharp C_p-λ characteristic and the other with a wide characteristic. Using those curves, numerical and simulation programs where written to obtain the values of K_ω and K_v along a constant power line for different wind velocities. The results obtained shown that in order to control the captured power by changing the rotational speed, the narrow blade characteristic is more suitable because with this sort of blade small changes in the rotational speed are necessary to reduce the captured power. However for below rated wind speed (BRWS) operation the control system should track the optimal value of λ as exact as possible because small variation from the optimal λ produces large reductions on the captured power.

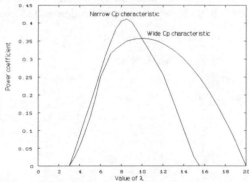

Figure 2.- Two different C_p-λ characteristics

3 CONTROL SYSTEM PROPOSED

Control system appropriates for power control by driving the machine to the aerodynamic stall zone has not been well investigated. Thiringer in [3] is the only researcher who has published a real implementation using this sort of control system in a real wind turbine of 20kW. Thiringer used wind speed estimation, based in a non linear function whose inputs are the aerodynamical torque on the blades and the rotational speed. However, wind speed estimation is not simple to obtain because for a given aerodynamical torque and a given rotational speed there are two possible wind speeds and it could be necessary that an external anemometer should be used in order to decide which wind speed is more representative of the effective wind speed through the blades. This fact is completely ignored in Thiringers and Linders [3] and no information about how to overcome this situation is discussed in that publication.

The torque observer implemented by Thiringer as well as the ones proposed and simulated in Connors and Leithead [2][4], and Ekelund [6], uses power measurement or torque transducers. In the presented paper the implementation is done using only a speed transducer because the electrical torque is assumed to be equal to the demanded electrical torque. The observer implemented for the mechanical torque is based on the equations:

$$\hat{T}_m = \frac{K_f z}{z^2 + a_p z + b_p}\left[T_e^* + \frac{J}{T}\omega_r(z-1)\right]$$

$$(11)$$

Where T_e^* is the demanded electrical torque and ω is the rotational speed. Equation (11) is the final transference function of the observer which has been designed using discrete state space techniques. This approach makes the implementation very simple because torque or power transducers are expensive, introduces further complication in the hardware as problem with electrical noise, extra delays produced by filters etc. This approach for the torque observer have been experimentally reported in Cárdenas et al [5] and the correlation between the demanded electrical torque and the real one for vector controlled induction machines and switched reluctance generator have been excellent.

The control strategy for the power limitation is based in that in order to keep the generated power constant for ARWS condition, the rotational speed for steady state conditions should be equal to:

$$\omega_{ref} = \frac{P_{max}}{T_m} \qquad \Delta\omega_{ref} = -\frac{P_{max}}{T_{mp}^2}\Delta T_m$$

$$(12)$$

Which means basicaly that when the power increases over the limit, ω the speed of the wind turbine has to be decreased in order to reduce the captured power.

One of the main complication with this sort of control system is that when the wind turbine machine is driven to the stall area there is a point where the open loop transfer function for the plant is unstable. This is because the variation of the mechanical torque with the

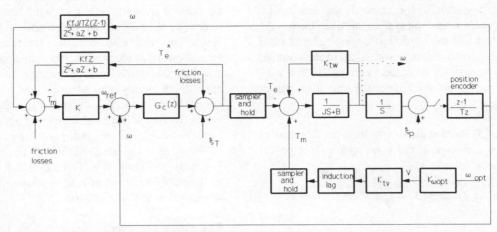

Figure 3.- Small signal model of the control system.

rotational speed becomes negative. If a stiff shaft is used then the following equation can be applied to the drive train:

$$T_m = J\frac{d\omega}{dt} + B\omega + T_e \Rightarrow$$

$$\frac{\partial T_m}{\partial V}\Delta V + \frac{\partial T_m}{\partial \omega}\Delta\omega = J\frac{d\Delta\omega}{dt} + B\Delta\omega + \Delta T_e \quad (13)$$

Using (12) and (13) the small signal model of the control system is shown in figure Figure 3. The terms $K_{t\omega}$ and K_{tv} are defined as the variation of the mechanical torque with the rotational speed and the variation of the mechanical torque with the wind velocity respectively. Table 1 shows the variation of $K_{t\omega}$ respect to the wind speed.

Table I.- Variation of $K_{t\omega}$ with wind speed.

Wind speed ms^{-1}	$K_{t\omega}$
10	0.302
13	-1.54
20	-2.37

To design a controller for this application is very difficult. This is because the system is non linear with large changes in the location of the poles and zeros produces by changes in the operating point with different wind speed. Also the system is non minimum phase and the value of the gain of the controller should be chosen carefully because a controller design which produces a good response for a given wind speed may produce an unstable or poorly damped response for others. The approach for the designing of the control system in this application has been to design a controller for an operation point and to test if this controller has an appropriate response for very high and very low wind speed. The design of a PI controller has been done using the root locus method.

4 EXPERIMENTAL RESULTS

The control system explained in section 2 have been implemented in the experimental system of Figure 4. This rig has been designed for testing the performance for several control strategies for below and above rated wind speed.

The physical emulation of the wind turbine is performed by a converter fed DC machine operating under control loop torque control. Real time wind samples are passed from a PC to a microprocessor which computes the turbine shaft torque via a real time simulation of the blade dynamics. This torque forms the torque demand to the DC drive. A position encoder of 720 samples per revolution is connected to the motor-generator shaft.

A switched reluctance machine has been used because of its excellent characteristic in terms of efficiency, robustness and control simplicity which make it competitive for wind energy applications (Cárdenas et al [7]). The switched reluctance generator is controlled via a look up table which maps the demanded electrical

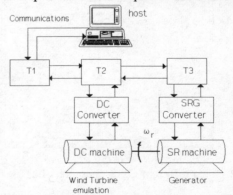

Figure 4.- Block diagram of the rig implemented.

torque and shaft speed into the switching angles θ_{on} and θ_{off} (positions to energise and de-energize the machine phases) that produce the desired torque. More information about the control of switched reluctance machines can be found elsewhere (Cárdenas et al [7]). Two T800 Transputers parallel processors share the control tasks as shown in Figure 4. These are excellent processors for real time development work particulary as they automatically provide for interactive communications between the user PC and all the system control variables. There is also another transputer in the rig for supervisory tasks.

For all the experimental results, the natural frequency of the observer is $\omega_o \approx 4\text{rads}^{-1}$ which although quite conservative is seen to give a good performance. The shafted nominal speed is 1500rpm, with a nominal torque of 47Nm and a rated power of 5.7kW. The controller has been designed to give a natural frequency of about 1rads^{-1} at $\approx 17\text{ms}^{-1}$ of wind speed. The nominal wind speed for this system is $\approx 9.8\text{ms}^{-1}$. Only the narrow blade characteristic of Figure 2 has been implemented in the control system of the DC machine. In the first part of this experimental work, the wind turbine control system has been tested using wind steps applied in different operating condition. The first wind step is applied when the machine is operating below rated wind speed and the rotational speed is near the nominal value. According to (10). The zone near the nominal rotational speed is very susceptible to changes in the wind speed. This is shown in Figure 5 where a wind step of 9 to 10.5ms^{-1} produces a large peak of power. The power change is considerably large and it can be controlled only with a large speed reduction from 1360 to 1137rpm after about 25sec.

Figure 6 shows another wind step but now from 12 to 13.5ms^{-1}. It is seen in this figure that the effect on the output power and rotational speed are completely negligible because the wind turbine is working near to

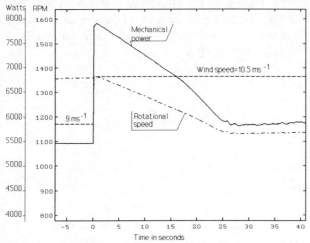

Figure 5.- Power and rotational speed for a wind step from 9 to 10.5 ms^{-1}.

the aerodynamic stall area. From the last figure it is also seen that the controller has a worse performances and that there is more amplification of the noise produce in

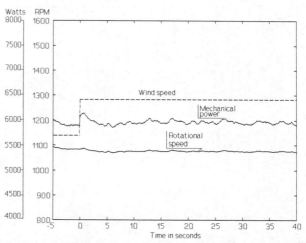

Figure 6.- Power and rotational speed for a wind step from 12 to 13.5 ms^{-1}

the speed encoder.

The next figure have been obtained using a rather smooth wind speed profile. Figure 7 shows the system working with a wind speed just above the nominal value with an average of 9.85 ms^{-1} and $\sigma=0.71$ ms^{-1} and with wind velocities in the range of 90 and 125% of the nominal value. It can be seen from this figure that the power peaks are rather high and that the rotational speed has relatively large variations.

Figure 8 shows a wind speed profile which is more turbulent. Again for the wind velocities near the nominal value (A in Figure 8), the power peaks are rather high and in the order of 140% of the nominal value, this power peaks are produced by wind velocities of about 130% of the nominal value. In the region B, the wind speed increases up to 155% the nominal value, however the power impact are just above 125%. This is because the rotational speed is much slower and near the 80% of the nominal value when the gust occurs and the machine is near the stall area where the variation of the captured power respect to the wind speed is not very high.

CONCLUSIONS

A formal mathematical analysis of the non linear dynamics and control system for the aerodynamic stall has been presented. The analysis is based on the control objective of constant power output.

The main difficulty associated with the control system designed in this chapter is that the transfer function produced by the interaction plant-observer has poles and zeros which changes noticeably its position in the z plane, making it very difficult to design a fixed control system which can produce a good response in all the possible operation condition.

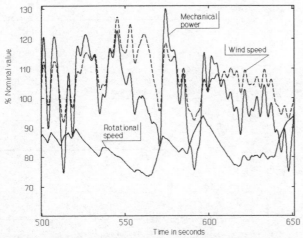

Figure 7.- Experimental result for low turbulence wind speed.

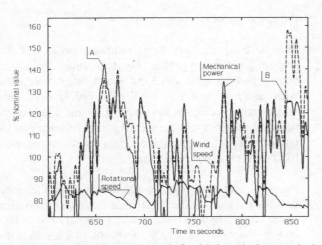

Figure 8.- Experimental result for high turbulence wind speed.

Simulation results have shown the suitability of blades having a narrow C_p-λ characteristics for the control system which used power limitation by aerodynamic stall. This is because with this sort of blades the necessary variation for the rotational speed and the overloading of the electrical torque are smaller than the values necessary with a broad blade. However, in term of power quality and tracking of the optimal trajectory for below rated wind speed conditions, the narrow blade characteristic has not advantages when compared to a traditional blade.

The non linear characteristic of this control system together with the variation of the poles and zeros of the transfer function , are matters for further research and possibly to use more complicated techniques of adaptive control in order to have an uniform and good performance in any wind speed condition.

ACKNOWLEDGMENTS

This work has been financially supported by the University of Magallanes Chile and the Govermment of Chile.

REFERENCES

[1] Goodfellow,D., Smith, G.A., 1986, "Control Strategy for variable speed wind energy recovery", Proceeding of 8th BWEA conf., pp 219-228.

[2] B. Connor and W. E. Leithead, 1994, "Control strategies for variable speed stall regulated wind turbines", 5th European Wind Energy Conference, pp 420-424.

[3] Thiringer, T., Jan Linders, J., 1993, " Control by Variable Speed of a Fixed-Pitch Wind Turbine Operating in a Wide Speed Range", IEEE TR. on Energy Conversion, Vol 8, No. 3, pp 520-526.

[4] B. Connor, W. E. Leithead, 1993, "Investigation of a fundamental trade-off in tracking the Cp max curve of a variable speed wind turbine", British wind energy conference, pp 313-319.

[5] R. Cardenas, G. M. Asher, W. F. Ray, 1995, "Optimal speed tracking in variable speed wind turbines using direct speed control and a torque observer", submitted for publication IEE proc-C.

[6] Thommy Ekelund, 1994, "Speed control of wind turbines in the stall region", Proc. 3rd IEEE conference on control application, pp 55-59.

[7] Cardenas R., Ray, W.F.,Asher, G.M., 1995, "Transputer based control of a switched reluctance generator for wind energy applications", EPE conference, Seville.

PERMANENT MAGNET GENERATORS FOR WIND POWER INDUSTRY: AN OVERALL COMPARISON WITH TRADITIONAL GENERATORS

N. Bianchi and **A. Lorenzoni**

University of Padova, ITALY

Abstract. A comparison between wounded rotor (WR) salient pole and superficial permanent magnet (SPM) synchronous generators for wind power industry is reported. A comprehensive analysis of all the design implications and manufacturing and operating costs has been carried out for both WR and SPM technologies, demonstrating the convenience of the innovative SPM machine design.

INTRODUCTION

The growing interest in renewable energy increases the scope for the improvement of energy conversion technologies as shown by Johansson [1]. Various energy sources and different electrical machines are currently studied for achieving high performance conversion. The dramatic improvement of performance and cost reduction for power electronics have made synchronous generator in diffused small size plant attractive as well as the induction one, which was preferred for its simplicity, see Hunt [2] and Murthy et al. [3].

The work aims at illustrating the optimal design of the generators for Wind Energy Converters (WEC) with particular interest in the innovative Superficial Permanent Magnet (SPM) synchronous generators.

A detailed analysis of the design criteria of traditional synchronous Wounded Rotor (WR) and innovative SPM generators is carried out, showing that the innovative generator design is less bounded than the WR one. In fact, different machine forms can be adopted with the purpose to decrease the operating costs. Moreover, the SPM generator reaches a higher efficiency because of the absence of excitation winding and corresponding Joule losses. In addition, it is shown that even if the regulation of magnetic field is not allowable, a SPM generator is characterised by high power factor and efficiency in full as well as in partial load operation.

A technical-economic comparison is also performed among them by evaluating their performance in specific wind conditions. A comprehensive analysis of all the investment, operating and maintenance (O&M) costs has been carried out for both the WR and SPM technologies. The economic superiority of the SPM technology is thus shown based on data relevant to a specific site in southern Italy, where many projects are under way (435 MW), are reported for a medium size generator.

At last, a tool to individuate the optimal size of the WEC generator is illustrated, based on the knowledge of the annual wind speed duration curve and the WEC cost function.

DESIGN CONSIDERATIONS FOR WR AND SPM SYNCHRONOUS GENERATOR

This section points out that, in comparison with the conventional WR salient pole generator, a SPM synchronous machine has reduced boundaries in the geometry, so that several solutions are possible. It can be built in different sizes, characterised by suitable length-to-diameter ratios. Moreover, it is not a difficult task to increase the number of pole pairs if a operating improvement can be achieved. In this section a comparison between the two generators is carried out, showing the key difference of design solutions.

General design equations

In designing synchronous machines, only the fundamental waveforms of the airgap flux density $B_g(\vartheta)$ and of the stator electric loading (linear current density) $K_s(\vartheta)$ are considered. Thus, the rated power developed at the airgap can be written as

$$P_r = \omega_m T = \omega_m \frac{\pi}{4} \hat{B}_g \hat{K}_s D^2 L \cos\gamma \qquad (1)$$

where T is the rated torque, D and L are the bore diameter and the stack length respectively, and γ is the angle between the symmetry axis of the airgap flux density distribution and the axis of stator electric loading distribution. At last, ω_m is the mechanical speed, related with the electrical speed as $\omega_m = \omega/p$ where p is the number of pole pairs.

The comparison between the two generators is carried out with constant peak values of the airgap flux density B_g and of the stator electric loading K_s. For a medium size generator, a suitable choice of airgap flux density is $B_g = 0.8 \div 1$ T. This flux density value can be obtained with a field winding in a WR generator or with a rare earth PM, such as the Neodymium-Iron-Boron (NdFeB) magnetic materials, which are characterised by a residual flux density $B_r = 1 \div 1.2$ T.

If the generator power P_r, the electric loading K_s and the airgap flux density B_g are constant, by using (1) with a fixed L/D ratio, the generator bore diameter D can be valued proportional to

$$D \equiv \left(\frac{\omega}{p}\frac{L}{D}\right)^{-\frac{1}{3}} \qquad (2)$$

Opportunities and Advances in International Power Generation, 18–20th March 1996,
Conference Publication No. 419, © IEE, 1996

On the contrary, the stack length is proportional to

$$L \equiv \left(\frac{\omega}{p}\right)^{-\frac{1}{3}}\left(\frac{L}{D}\right)^{\frac{2}{3}} \qquad (3)$$

Because of the assumed hypothesis of constant P_r, K_s and B_g, the airgap volume does not depend on the length-to-diameter ratio, while it is proportional to the machine torque. If the power is given and the mechanical speed decreases, a higher torque is required, so that a bigger size of the generator is unavoidable. On the contrary, the airgap surface is connected to the L/D ratio and the pole pairs number as

$$\pi D L \equiv \left(\frac{\omega}{p}\right)^{-\frac{2}{3}}\left(\frac{L}{D}\right)^{\frac{1}{3}} \qquad (4)$$

while the number of conductors per phase, with an established winding factor, can be written as

$$N \equiv \frac{1}{DL}\frac{p}{\omega} = \left(\frac{\omega}{p}\frac{L}{D}\right)^{-\frac{1}{3}} \qquad (5)$$

By comparing (5) and (4), one can note that the conductors number decreases with an increase of the L/D and of the working frequency, or when the number of pole pairs is reduced. In fact, with a higher L/D, the conductor active length increases, while the end-winding lengths decrease because of the smaller bore diameter.

Effect of the length-to-diameter ratio

In designing a WR generator the length-to-diameter ratio L/D is chosen quite small, to facilitate the field winding building and the correspondent cooling system. Typical values can be L/D=0.5÷0.7. On the contrary, in the SPM generator the field winding is absent, so there are not boundaries in the rotor design and a higher length-to-diameter ratio can be chosen, i.e. L/D=1÷3. Moreover, a rotor cooling system is not necessary lacking rotor losses.

From (2) the bore diameter can be reduced to 0.5÷0.8 times the WR one in a SPM generator with a high L/D, while the stack length (3) increases. As mentioned above the airgap volume does not depend on L/D, while the airgap surface increases with it as (4) shows.

If a purpose of the design is to reduce the PM weight, that is the most expensive material in a SPM generator, an airgap surface as little as possible must be chosen. In fact, if constant airgap and PM length l_m are considered, the PM weight is proportional to the airgap surface. To reduce this surface corresponds to select a low L/D.

Another possible aim in the design is to cut the copper losses. By choosing the same current density, the SPM and the WR generators require an equal copper section S_c. Let the end-winding length be proportional to the bore diameter D by using the coefficient k_{ew}, the copper weight W_{Cu} can be written as

$$W_{Cu} = \gamma_{Cu} 3 N S_c (L + k_{ew} D) \qquad (6)$$

where γ_{Cu} is the copper weight density. By rearranging (2), (3) and (5) in (6), it can be noted that the copper weight is in relation with the L/D ratio as

$$W_{Cu} \equiv \left(\frac{\omega}{p}\right)^{-\frac{2}{3}}\left[\left(\frac{L}{D}\right)^{\frac{1}{3}} + k_{ew}\left(\frac{L}{D}\right)^{-\frac{2}{3}}\right] \qquad (7)$$

It assumes a minimum value when L/D=2k_{ew}. As k_{ew} can be estimated as 2.5/p, it is worth to observe that the minimum copper weight is achieved with L/D=5/p. Thus with the mentioned ratio, the SPM generator needs a copper volume equals to (75÷90)% of the WR one, that is (25÷10)% less. With the same stator current density J_s, the copper losses

$$p_{Cu} = 3RI^2 = \frac{\rho_{Cu}}{\gamma_{Cu}} W_{Cu} J_s^2 \qquad (8)$$

are proportionally reduced.

On the contrary, the necessary stator iron volume and the correspondent losses increase with a high L/D. Since in the two typologies of generators identical values of stator electric loading K_s and current density J_s are supposed, it is simple to verify that the slot height is practically the same in the two cases. In addition, with identical iron flux densities, the stator back iron height is proportional to the ratio D/p; let α be the proportionality coefficient. Nevertheless, by observing (2) and (3), when the diameter decreases the stack length increases in a squared way, so that the stator iron volume increases with L/D. By adopting a constant coefficient k to consider that the slot height does not change with L/D, the stator iron weight can be expressed as

$$W_{slot} \equiv k\left(\frac{\omega}{p}\right)^{-\frac{1}{3}}\left(\frac{L}{D}\right)^{\frac{2}{3}}\left[\left(\frac{\omega}{p}\frac{L}{D}\right)^{-\frac{1}{3}} + k\right] \qquad (9)$$

$$W_{back} \equiv \frac{1}{p}\left(\frac{\omega}{p}\right)^{-\frac{2}{3}}\left(\frac{L}{D}\right)^{\frac{1}{3}}\left[\left(\frac{\omega}{p}\frac{L}{D}\right)^{-\frac{1}{3}} + 2k + \frac{\alpha}{p}\left(\frac{\omega}{p}\frac{L}{D}\right)^{-\frac{1}{3}}\right] \qquad (10)$$

Higher L/D ratios in a SPM generator generally lead to higher stator iron volume and iron losses. By using the previously defined L/D ratios, for a SPM generator the stator iron volume increasing can be valued in the region of (20÷50)%.

Effect of the number of pole pairs

An improvement in the design can be realised by choosing a suitable number of pole pairs. Rearranging this number is not a difficult task with a SPM generator. In fact, while a high number of pole pairs implies an impracticable building of the WR generator because of the salient poles, the operation results very simple in the SPM case. In changing the number of pole pairs two cases must be considered, with different results:

i. If the working frequency is kept constant, when the generator must be connected to the grid, a low number of poles should be chosen. In fact, with a higher pole pairs number, the mechanical speed is reduced, so that the machine torque increases, see (1). The generator size and the materials weight become higher, thus the machine losses increase, causing a lower efficiency. As a conclusion, a high pole pairs number gives rise to a bigger, expensive and inefficient generator.

Nevertheless, an interesting solution in this direction is to use a direct-coupled generator, as in Spooner and Williamson [4-5]. The attractiveness of this solution is that the gearbox is absent and in particular this synchronous SPM generator is suitable for low-speed operation. The disadvantages of a gearless direct-coupled machine are large radial diameter (2) and the high weight of the generator; the advantage of a gearless wind turbine is instead the reduced maintenance and the higher efficiency. Nevertheless, today a gearless wind turbine is generally more expensive than a conventional one (see Table 1).

ii. In the second case, a constant mechanical speed is considered, while the working frequency is proportional to the pole pairs number. This is when the machine operates at frequency different from the grid one and a frequency converter is necessarily used. From (1), when ω_m is constant, machine torque T and size D^2L do not change; thus, when the pole pairs number increases, end-winding length and back iron height decrease. Copper weight (7) and iron weight (9)-(10) decrease, as well as the copper losses (8). The iron losses generally become a little higher because of the higher working frequency. As an alternative, a lower iron flux density can be chosen to obtain a high efficiency. At last, given the mechanical speed, a cheap and efficient solution should be a SPM generator with both high number of pole pairs and working frequency.

Reactance and armature reaction

The values of the leakage reactances will be now analysed. When the L/D increases, the end-winding reactance is proportionally reduced, while the slot reactance remains almost the same. It is described as follows

$$X_{slot} \equiv \omega N^2 L \equiv p \qquad (11)$$

$$X_{ew} \equiv \frac{\omega}{p^3} N^2 D \equiv \frac{1}{p^2} \left(\frac{L}{D} \right)^{-1} \qquad (12)$$

Then, in a SPM generator designed with a higher L/D the total leakage reactance is lower than in a WR generator. Moreover, in the SPM generator, the magnetising reactance is greatly reduced. In fact it decreases when L/D increases; besides it greatly depends on the distance between the stator and rotor iron. This distance takes into account the airgap g as well as the PM length l_m, so that the magnetising reactance becomes

$$X_{magn} \equiv \omega \frac{N^2}{p^2} \frac{DL}{g + l_m / \mu_r} \equiv \frac{1}{p} \left(\frac{\omega}{p} \frac{L}{D} \right)^{-\frac{1}{3}} \frac{1}{1 + l_m / g\mu_r} \qquad (13)$$

Since the PM relative differential permeability $\mu_r \approx 1$ in the working region and it is generally $l_m/g = (3 \div 6)$, the SPM reactance is $(20 \div 25)\%$ times the typical d-axis reactance of a salient pole WR generator. Then, the torque angle δ of a SPM generator, between the full load voltage and the back e.m.f. E, is very little in comparison with a salient pole generator.

In a SPM generator the back e.m.f. can be chosen to obtain the lowest current in the full load working point, so that a unitary power factor is achieved. Because of the small reactances, the back e.m.f. is close to the output voltage, thus when a SPM generator works at reduced power, the power factor assumes high value as well. Fig.1 shows the power factor of a SPM generator as a function of the torque angle. Fig.2 shows the vector diagram of the electric quantities in p.u. value of a SPM generator at full and at half load. At full load, the nominal torque angle δ_N to achieve a unitary power factor $\cos\varphi = 1$ is assumed, while at half load the power factor angle φ is still very small.

It is worth to note that although the power factor regulation is not indispensable if the generator works with a static converter, the power factor gives an index of the required current with an equal value of the produced power.

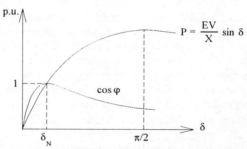

Figure 1: SPM power and power factor as functions of torque angle δ

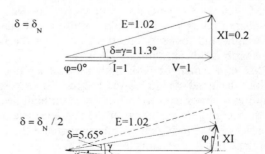

Figure 2: Example of SPM generator full load and half load vector diagram

On the contrary, the back e.m.f. E of the WR generator can be arranged by the field winding, so that the current phase can be regulated with respect to the voltage. However, high excitation power is needed because of the magnetising reactances and the intense armature reaction. A higher difference between no load and full load conditions can be observed. For example, the required nominal power in the field winding can be evaluated as $p_{field}=(2÷4)\%$ P_r for a 100kVA generator, which is a typical size for wind power applications. This power proportionally reduces the generator efficiency.

By using a SPM generator, the rotor winding and the correspondent control are removed, so the excitation power is reduced to zero, and the total efficiency increases of 2÷4 per cent. The high cost of the rare earth PM can be quickly neutralised by the minor losses during the machine operation as shown below.

Overload capability

The capability of the machine to resist torque higher than rated for short periods during gusts must be verified. Since SPM generator works with a little torque angle δ_N, a temporary overload is possible, as Fig.1 shows. This overload can be estimated as $1/\sin(\delta_N)$ in p.u.. In the overload transitory, the current assumes very high values: it is essential to verify that these values do not irreversibly demagnetise the PM.

GENERATOR DESIGN EXAMPLES

This section shows some generator design solutions for wind-power industry. Designs have been proposed for a turbine of 18 m swept diameter, rated power of 100 kVA and nominal blade tip speed supposed 60 m/s, so that the turbine angular speed results in 6.68 rad/s.

Table 1 shows four different generator geometries, and their main dimensions, weight, losses and performance are pointed out. WR represents a conventional salient pole generator with electrical frequency 50 Hz, while the remaining designs are based on surface mounted NdFeB magnet generators, where the PM is characterised by a residual flux density of 1.1 T. The L/D ratio is chosen to minimise the copper weight (7) and the copper losses in all cases.

SPM1 refers to a generator with speed, frequency and number of pole pairs equal to the WR one. A reduction of weight and losses is observed, although the use of rare earth magnets entails a higher cost.

SPM2 is a generator with higher frequency and pole pairs, designed for a lower torque: it necessarily requires a power converter. It is noted an high decrease of the weights, in particular of the PM one. The higher iron losses due to higher frequency increase the losses, but the efficiency remains high.

At last, SPM3 refers to a gearless direct-coupled generator with angular speed equal to that of the turbine shaft. The number of pole pairs has been chosen to give the same electrical frequency of WR. In this solution, weight and losses are higher. The generator efficiency decreases, becoming lower than that of the

WR generator, although the system efficiency can be higher thanks to the gearbox absence, being the gearbox efficiency 2% per stage and its cost some 7% of the total investment cost, as reported in Terrinoni and Ferrari [6]. Nevertheless, this solution is actually very expensive, as it requires a large PM volume. Moreover, when the generator rated power increases, the dimensions become hardly manageable.

TABLE 1 - Parameters of generators (P_r=100kVA)

		WR	SPM1	SPM2	SPM3
ω_m	rad/s	78.5	78.5	104.7	6.68
2p	-	8	8	16	94
D	m	0.463	0.310	0.356	0.96
L	m	0.185	0.390	0.222	0.48
OD	m	0.626	0.444	0.463	1.05
W_{Cu}	kg	90	55	35	150
W_{Fe}	kg	360	315	150	605
W_{PM}	kg	-	23	15	90
W_{tot}	kg	450	393	200	845
losses	W	6900	4170	4920	11000
effic.	%	93.6	96.0	95.3	90.2
revenue	$	12205	12660	12790	11985
ROI	%	10.31	10.34	10.61	9.32

Fig.3 shows the different size of the generators presented in Table 1. As explained above, it is worth to note that SPM generators can be designed in different forms and are generally more convenient than conventional WR.

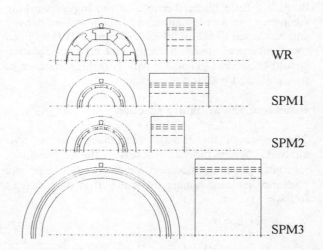

Figure 3: Dimensional comparison of studied wind-power generators.

ECONOMIC CONVENIENCE OF THE SPM GENERATOR

In order to compare effectively the considered generators, their operation has been simulated in the same atmospheric conditions in a specific site in Southern Italy, where a wind farm has been installed.

Wind speed data measured in Frosolone (Isernia province) were available and have been taken as reference conditions: the wind speed duration curve is reported in Fig.6(a).

The hypothesis is made to fit a WEC with WR and then with SPM generators and to compare their economic performance. The characteristic of the reference WEC are reported in Table 2 and are derived from the machines actually on the market [7]; the power output P is given by

$$P = 1/2 \, C_p \, \eta_{mc} \, \eta_{el} \, \rho \, \pi \, r^2 \, v_w^3 \qquad (14)$$

where C_p is the coefficient of performance, function of the wind speed when variable pitch blades are adopted (its value can be deduced from Fig.4); η_{mc} and η_{el} are the mechanical and electrical efficiencies: η_{mc} is constant at 95%, except in SPM3 which has η_{mc}=98%, while η_{el} is evinced from Table 1 for both WR and SPM generators; ρ is the air density (1.224 kg/m^3), r is the blade length (8.92 m), v_w the wind speed (m/s).

TABLE 2 - Reference WEC parameters

swept area	250 m^2
v_w cut in	3 m/s
v_w cut off	26 m/s
total inv. cost WR (100 kW)	1184 US$/kW
total inv. cost SPM1 (100 kW)	1224 US$/kW
O&M cost	2.0% inv. cost WR
	1.5% SPM1 and 2,
	1.2% SPM3
WEC life	15 years
real discount rate	5%
electricity unit price p_u	0.108 US$/kWh.

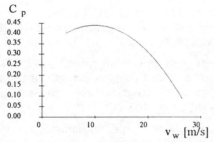

Figure 4: C_p as a function of the wind speed

The produced energy is priced like the wind energy sold to the Italian electric power grid in 1995, i.e. 173.5 ItLire/kWh=0.108 US$/kWh. The economic profitability of the investment has been evaluated when the WEC is equipped with a WR and a SPM generators in *ceteris paribus* conditions, i.e. having the machine the same blades and mechanical efficiency. Thus the comparison has been carried out on the revenues from energy produced and on the Return on Investment (ROI), as shown in the last two rows of Table 1. The convenience of the SPM generators is proved by the higher revenue as well as by the higher revenue to investment costs ratio (ROI); although the yearly profitability gain is small, it is considerable in the

whole WEC life. With this generators size, SPM3 performs worse than others, due to lower η_{el} and the incidence of the PM cost, in spite of the cost reduction (both investment and O&M) and the η_{mc} increase.

The analysis is focused on the average performing SPM1 generator compared to the WR. Although the SPM generator is more expensive, it has two main advantages on WR: a higher efficiency and a lower O&M cost. Moreover the cost of the generator accounts for just some 5% of the total investment and only partially compromises its profitability.

The efficiency gain is straight reflected in an increase of the annual electricity yield, while the reduction of O&M cost is estimated in the range of 25%. In Fig.5 the total yearly revenue, obtained from the difference between proceeds and costs, is shown in the two cases for different sizes of the installed generator. As can be easily seen in Fig.5 the SPM1 is always convenient compared to the WR.

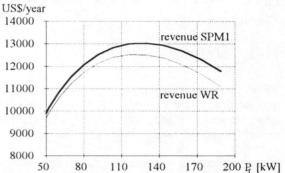

Figure 5: Profitability of WEC with different generators.

CHOICE OF THE OPTIMAL GENERATOR SIZE

Although it is not thinkable to change the size of the generator in each application, it is interesting for a given wind turbine to identify the optimal size for its generator in wind conditions typical for a geographic area (the Southern Appennines mountains in our case). According to the concepts discussed above the total WEC investment cost C has been written as a function of the ratio p_r between the rated power P_r and the reference 100kVA generator for both WR and SPM

$$C_{WR} = C_{Cus}p_r^{0.25} + C_{Fe}p_r^{0.75} + C_{Cur}p_r^{0.5} + C_{pe}p_r + C_{ot}p_r^{0.85}$$

$$C_{SPM} = C_{Cus}p_r^{0.25} + C_{Fe}p_r^{0.75} + C_{PM}p_r^{0.5} + C_{pe}p_r + C_{ot}p_r^{0.85}$$

$$(15)$$

C_{Cus}, C_{Cur}, C_{Fe} and C_{PM} are the costs of the materials (stator and rotor copper, iron and PM), C_{pe} is the cost of power electronics and C_{ot} is the factor that keeps into account all the other investment costs, supposed identical in the two cases. These functions have been designed to be as close as possible to the trend of the costs; nevertheless different expressions could be conceived if they would better describe the costs observed in practice.

54

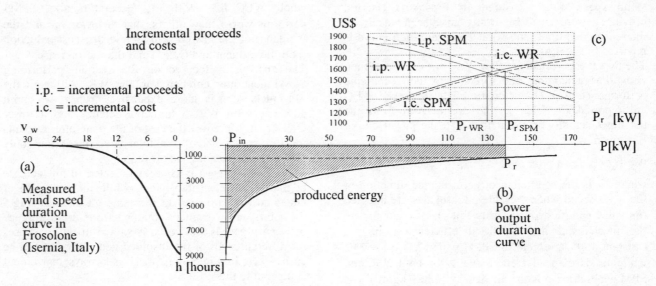

Figure 6: Tool for the choice of the optimal size of the generators and nominal wind speed

The proceeds R instead are given by

$$R = p_u \int_{P_{in}}^{P_r} h(P)dP \qquad (16)$$

where p_u is the electricity unit price, P_{in} is the power generated at v_w cut in, and $h(P)$ is the hour vs. power curve shown in Fig.6(b). The integral in (16) is thus the energy generated under the power duration curve in Fig.6(b).

Demonstrated the improvement achievable with SPM generators at all rated wind speeds, a tool is shown to evaluate the optimal size of the generator and the optimal rated v_w when the wind speed duration curve and the wind turbine are given. In fact, it is essential to evaluate the sensitivity of the profitability of the investment to the size of the generator: the power rate P_r can be augmented as far as the yearly cost C_y (investment plus O&M cost) growth is covered by the growth of the produced energy. The economic theory says the optimum is reached when the marginal investment cost dC_y/dP_r equals to the marginal proceeds dR/dP_r; the optimal size is thus given by the condition

$$\frac{dR}{dP_r} - \frac{dC_y}{dP_r} = 0 \qquad (17)$$

The trend of the marginal costs and proceeds is shown in Fig.6(c) to calculate the optimal generator size. It is interesting to note that, given the same blades and wind conditions, the optimal size is different for the SPM and the WR case, as a consequence of the different efficiency and costs. When P_r is calculated, the number of operating hours at the rated power and the rated v_w are immediately found from the curves in Fig.6(a). The characteristics of the WEC for the optimal profitability are thus all known.

Under the considered conditions (a fundamental role is played by the price of electricity) we have found the economic best for a nominal 100 kW WEC is to be

fitted with a 135 kW generator in the case of SPM1 generator and 126 kW when a WR would be preferred. Very different results are obtained with others wind conditions and electricity prices.

List of References

1 T.Johansson, H.Kelly, A.K.N.Reddy, R.H. Williams (editors), 1993: "Renewable Energy, sources for fuels and electricity", Island Press Washington D.C. & Earthscan London

2 V.D.Hunt, Wind Power - A handbook on wind energy conversion systems, 1981, Van Nostrand Reinhold Company

3 S.S.Murthy, O.P.Malik, A.K.Tandon, "Analysis of self-excited induction generators", 1982, IEE Proc., vol.129, Pt.C, no.6

4 E.Spooner and A.C.Williamson, 1992, "The feasibility of direct-coupled permanent-magnet generators for wind power applications", SPEEDAM Conf., 105-111

5 E.Spooner and A.C.Williamson, 1992, "Permanent-magnet generators for wind power applications", ICEM Conf., 1048-1052

6 L.Terrinoni and G.Ferrari, "Costs and benefits of a wind farm" (in italian), 1994, Energia Ambiente e Innovazione, n.6, June

7 "Product Guide - Wind Turbines", 1990, Modern Power Systems, October

Address of authors

Dr. Nicola Bianchi, Dr. Arturo Lorenzoni
Department of Electrical Engineering, University of Padova, Padova (ITALY). Tel. ++39.49.827.7574; FAX: ++39.49.827.7599
e-mail LORART@maya.dei.unipd.it

ELECTROMAGNETIC ANALYSIS OF A LOW-SPEED PERMANENT-MAGNET WIND GENERATOR

P. Lampola, J. Perho

Helsinki University of Technology, Finland

ABSTRACT

A directly driven permanent-magnet wind generator was designed. The rated power of the generator was 500 kW and the synchronous rotational speed 40 rpm. The electrical performance of the generator was calculated by the finite element method. The generator will be used with a frequency converter in order to allow variable speed operation.

INTRODUCTION

Wind power technology has been developed much during the last decade. The real cost of energy from wind turbines is falling dramatically. Nowadays more than 3700 MW wind power capacity has been installed world-wide. The rotor of a wind turbine rotates typically at a speed of 20 – 100 rpm. In conventional wind power plants the generator is coupled to the turbine via a gear so that the generator can rotate at a speed of 1000 or 1500 rpm. However, the gearbox brings weight, generates noise, demands regular maintenance and increases losses. In a directly driven, low-speed wind generator no gears are needed and the above-mentioned disadvantages can be avoided, Fig. 1.

A low-speed generator can be connected to the grid via a frequency converter. This choise has benefits to the generator design. The low speed and torque ripple can together cause problems, if the generator is directly connected to the grid. Using the converter the generator frequency can be lower than the frequency of the network. This means the generator can have reduced number of poles than at network frequency. The converter-generator can be used at variable-speed operation according to the wind.

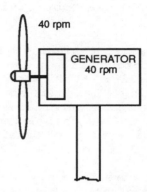

Figure 1. Directly driven wind power plant.

The aim of this study is to analyse a directly driven permanent-magnet wind generator.

METHOD OF ANALYSIS

The electromagnetic optimisation of the generator was done by the finite element method. The program used was developed at the Helsinki University of Technology in the Laboratory of Electromechanics by Arkkio (1) and Väänänen (2). The calculation of the operating characteristics of the generator was based on a time-stepping, finite element analysis of the magnetic field. The field was assumed to be two-dimensional. The losses of the machine were calculated with a sinusoidal supply.

The laminated stator core was modelled as non-conducting, magnetically non-linear medium. The rotor core, which was made of massive steel, was modelled as conducting, magnetically non-linear medium. The permanent magnets were modelled as conducting material.

PERMANENT-MAGNET GENERATORS

Construction of the generator

A directly driven permanent-magnet wind generator was designed. The rated power of the generator was 500 kW and the synchronous rotational

Opportunities and Advances in International Power Generation, 18–20th March 1996,
Conference Publication No. 419, © IEE, 1996

speed 40 rpm. The simplest way to construct a rotor with a great number of poles was mounting the magnets onto the surface of the rotor yoke, Fig. 2. In this study it was necessary to use high-energy magnets such as NdFeB magnets to provide an acceptable flux density in the air gap. The remanence of the NdFeB permanent magnets used was $B_r = 1.14$ T and the coercivity $H_C = 850$ kA/m.

A fractional slot winding was used to get the torque ripple and cogging torque smaller than with integral slot winding, Lampola et al. (3). Very much magnet material would have to be used in the designed machine if the peak air-gap flux density should exceed 0.8 T, Lampola (4).

The stator core segments were made of 0.5 mm laminations. The maximum flux densities were 0.8 T and 1.7 T in the air-gap and stator tooth, respectively. The rotor yoke was a cylinder made of massive steel. The main parameters of the generator are given in Table 1 and the weights of the active material are shown in Table 2.

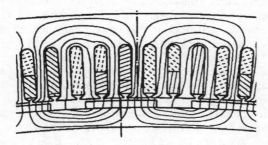

Figure 2. The cross-sectional geometry of the permanent-magnet generator. The magnetic field shown corresponds to the operation at the rated power. The difference between the lines is 5 mWb/m.

TABLE 1. Main parameters of the permanent-magnet generator.

Rated output [kW]	500
Rated voltage [V]	590
Rated current [A]	490
Rated frequency [Hz]	26.7
Connection	star
Stator outer diameter [mm]	2670
Stator inner diameter [mm]	2500
Rotor inner diameter [mm]	2440
Core length [mm]	520
Air-gap length [mm]	2.5
Magnet's height [mm]	7
Number of poles	80
Number of phases	3

TABLE 2. Active material weight of the permanent-magnet generator.

Material	Weight kg
Stator core (laminated)	1870
Rotor yoke (massive)	610
Stator winding (copper)	670
Magnets (NdFeB)	140
Total active weight	3290

Performance of the generator

The calculation of the operating characteristic of the machine was made by the finite element method. The waveforms of the winding voltage and cogging torque at no-load and torque at rated load are shown in Fig. 3. The waveforms of the induced voltage and cogging torque were calculated so that the stator winding was open, i.e. it was not connected to any external source. That means that the stator current was zero. The waveform of the torque was calculated as the generator is connected to a network the voltage of which is sinusoidal. The maximum air-gap torque was 2.4 times the rated air-gap torque.

a:

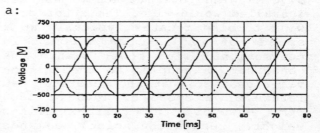

b:

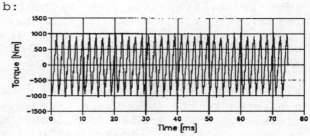

c:

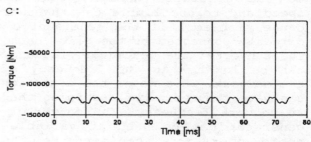

Figure 3. The waveforms of the winding voltage (a) and cogging torque (b) at no-load and torque (c) at rated load.

The cooling of the machine was analysed by a thermal network (3). The temperature rise of the stator winding and magnets was 80 °C and 20 °C, respectively. The generator was chosen to be asymmetrically cooled with radial cooling ducts and a flow rate of 1 m^3/s. The generator had constant speed external ventilator, which electrical power was 1 kW.

Efficiency of the generator

Most of the losses were concentrated in the stator winding, but rotor core and permanent magnets had very low losses, only 0.4 kW at rated load. The generator loss distribution under different loading conditions is shown in Fig. 4. The efficiency and power factor of the generator at rotational speeds of 20 - 40 rpm are shown in Fig. 5. The speed was 40 rpm at an output power of over 200 kW and 20 - 40 rpm at a power of 25 - 200 kW. The efficiency at rated load was 95.4 %.

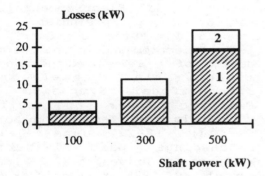

Figure 4. Loss distribution of the permanent-magnet generator. 1: stator resistive losses, 2: iron losses.

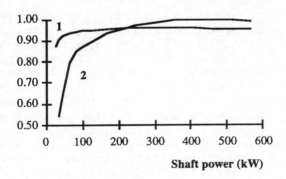

Figure 5. Efficiency (1) and power factor (2) of the variable speed permanent-magnet generator as a function of the shaft power.

WIND POWER PLANTS

The electromechanical system of the designed wind power plant consists of three main parts: turbine, low-speed generator and frequency converter. The generator was connected directly to the main shaft.

Efficiency of the wind power plant

Total losses consist of the losses of the generator, ventilator and converter. The losses of the generator was calculated by the finite element method. The converter losses was calculated by the analytical method, Grauers (5). The total converter losses can be described as a sum of the no-load losses of the inverter, the voltage drop losses of the diode rectifier and the thyristor inverter and the resistive losses of the rectifier, dc filter and inverter. The converter efficiency at rated load was 98 %. The total efficiency of the wind power plant is shown in Fig. 6.

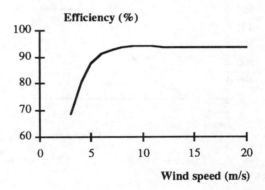

Figure 6. Efficiency of the wind power plant.

Energy production

Energy production of the wind power plant depends on the wind conditions. Three different places were chosen for mounting a wind power plant with the designed low-speed generator: inland (Jokioinen), coast (Isosaari) and offshore (Kalbådagrund). The average wind-speed distributions of the chosen places are shown in Fig. 7, Tammelin (6). The design criteria for the wind turbine were the minimum operational wind speed 3 m/s and the cut-out wind speed 20 m/s. The calculated output of the wind power plant is shown as a function of the wind speed in Fig. 8.

In inland the energy production was only 300 MWh, but in coast 1200 MWh and in offshore 1800 MWh. As a summary, the chosen places are collected into Table 3 with information of average wind speed, annual energy production and effective full load working hours.

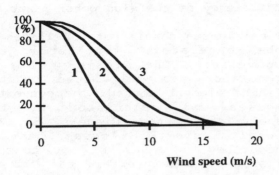

Figure 7. Distributions of average wind speed at Jokioinen (1), Isosaari (2) and Kalbådagrund (3).

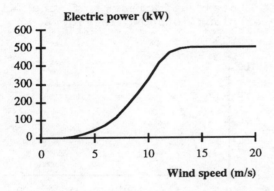

Figure 8. Electric output of the wind power plant.

TABLE 3.Average wind speeds, energy productions and full load hours.

	v_{av} m/s	E MWh/a	t_{fl} kWh/kW
Jokioinen	3.8	300	600
Isosaari	6.5	1200	2400
Kalbådagrund	8	1800	3600

CONCLUSIONS

A 500 kW, 40 rpm directly driven permanent-magnet wind generator was designed. The electrical performance of the generator was calculated by the finite element method. The excitation of the generator was made by NdFeB magnets mounted on the surface of the rotor yoke. The

diameter of the machine was 2.5 m, the length 0.5 m and the active weight 3290 kg. The efficiency of the generator at rated load was 95.4 %. The generator will be used with a frequency converter in order to allow variable speed operation.

A low-speed permanent-magnet synchronous generator would be a good solution for the construction of a directly driven wind generator.

REFERENCES

1. Arkkio, A. 1987. "Analysis of induction motors based on the numerical solution of the magnetic field and circuit equations." Helsinki, Finland: Acta Polytechnica Scandinavica, Electrical Engineering Series, Doctoral thesis, No. 59, 97 p.
2. Väänänen, J. 1992. "Calculation of electric and thermal properties of a permanent-magnet high-speed electrical motor." Espoo, Finland: Helsinki University of Technology, Laboratory of Electromechanics, Report 34, 93 p. In Finnish.
3. Lampola, P., Perho, J., Saari, J. 1995. "Electromagnetic and thermal design of a low-speed permanent magnet wind generator." *International Symposium on Electric Power Engineering (Stockholm Power Tech)*, Stockholm, Sweden, 18-22 June 1995, Proceedings, part Electrical Machines and Drives, p. 211-216.
4. Lampola, P. 1995. "Directly driven generators for wind power applications." *EWEA Special Topic Conference on The Economics of Wind Energy (EWEA 95)*, Helsinki, Finland, 5-7 September 1995, Proceedings, p. E4: 1-6.
5. Grauers, A. 1994. "Synchronous generator and frequency converter in wind turbine applications: system design and efficiency." Göteborg, Sweden, Chalmers University of Technology, Department of Electrical Machines and Power Electronics, Technical report 175 L, 125 p.
6. Tammelin, B. 1991. "Finland wind atlas." Helsinki, Finland, Finnish Meteorological institute. 355 p. In Finnish.

ELECTRICAL EQUIPMENT FOR A COMBINED WIND / PV ISOLATED GENERATING SYSTEM

F. Crescimbini;[1] F. Carricchi;[1] L. Solero;[1] B. J. Chalmers;[2] E. Spooner;[3] Wu Wei.[2]

1: University of Rome La Sapienza   Italy
2: UMIST   UK
3: University of Durham UK

ABSTRACT

Wind and solar photovoltaic generating systems have advanced rapidly in recent years, but both suffer from the intermittent nature of their resource. A secure stand-alone supply based on either requires a substantial energy store. Combining the two creates a supply with smaller storage requirement since the fluctuations have a statistical tendency to cancel.

The system developed for a 10 kW application comprises wind and solar collectors, each of 5 kW rating, with a lead-acid battery for storage and a 10 kW PWM inverter for the final output. The wind turbine generator is a 5 kW, 200 rpm, direct-drive, permanent-magnet, axial-flux machine based on the 'Torus' configuration. The three-phase output is rectified to form a variable voltage dc link. The power converter uses two dc-dc converters connected in series, each with a bypass diode which conducts continuously when the corresponding source is not available.

INTRODUCTION

There is a potentially vast world market for stand-alone power sources. In rural districts of the developing world, the energy consumption per capita is very low, the basic energy needs are for water pumping, electricity supplies to small hospitals, lighting, cooling and telecommunications. Often the cost of connection to the grid in remote locations cannot be justified.

Photovoltaic and wind power can meet these needs, but Either source alone provides an intermittent supply and energy storage is needed to deliver a reliable supply. However, these two sources are complementary since sunny days are usually calm and strong winds are often accompanied by cloud and may occur at night. A combined plant therefore has higher availability than either individual source and so needs less storage.

In a collaborative project supported by the EU JOULE II programme a 10 kW demonstration plant has been designed nominally for a site in southern Italy and prototype components have been constructed. The prototype design is simple and can continue to operate even if one of the input devices has failed.

SYSTEM DESCRIPTION

The prototype system has the following specification. However, the plant could easily be modified to provide a single-phase phase output or to utilise a different proportion of wind and solar energy according to local circumstances and needs.

Overall system
Rated output power   10 kW
          at 380 V, three phase, 50 Hz
Wind turbine
Rated electrical output power   5 kW
          at 150 V dc
Turbine diameter   5 m
Turbine speed   100 - 200 rpm
Available space for machinery   500 mm diameter
          Targets for the generator design are:
          low cost, high efficiency (90% or better).
Solar panel
Rated electrical output power   5 kW
          at 150 V dc

The prototype design is illustrated in Figure 1. The heart of the system is the double-input, single-output dc power conditioner.

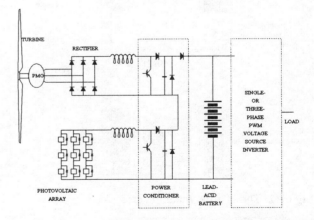

Figure 1   Overall plant electrical arrangement

PROTOTYPE COMPONENTS

PV Array

The PV array uses 14 parallel strings of eight series-connected standard modules. A module comprises 36 polycrystalline silicon cells housed in an aluminium frame with polyester back and glass front covers. The array voltage is 168 on open-circuit falling to 136 at the design power of 5 kW. The mean efficiency is 13%. For maximum energy capture, the array should be tilted from the horizontal and the angle should be varied though the year. For simplicity, provision is made to set

Opportunities and Advances in International Power Generation, 18–20th March 1996,
Conference Publication No. 419, © IEE, 1996

the angle at 20° during the summer and 60° in winter. It was calculated that acceptable energy production would be achieved.

Wind Turbine Generator

Design. The wind turbine section of the plant is to use a directly-coupled generator. In recent years a number of direct drive concepts have been proposed[1,2,3], or tried:

hydro-type synchronous generators with wound poles for a vertical-axis 4 MW turbine and for the Enercon 500 kW production machine

A small battery charging machine use an ironless-stator, axial-flux, permanent-magnet generator.

Proposals for direct-coupled grid-connected generators include transverse-flux and radial flux permanent-magnet machines.

The 'Torus' configuration[4,5,6] illustrated in Figures 2 and 3 is an axial-flux permanent-magnet machine using a slotless iron-cored stator and has been selected for the present application.

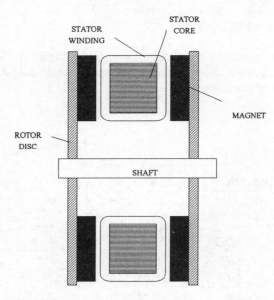

Figure 2 The 'Torus' machine cross section

Materials. Several types of permanent-magnet material have been used in electrical machines including Alnico metal alloys in pilot exciters for large turbine generators, ferrite for small low-cost motors and rare-earth alloys such as Nd-Fe-B used in compact high-performance motors. For minimum cost, ferrite would be preferred but in order to meet the small diameter restriction and the target for high efficiency, Nd-Fe-B is found to be necessary.

The specific costs of ferrite and Nd-Fe-B are approximately 5.0 and 100 ECU/kg respectively. Cutting to size adds additional cost to both.

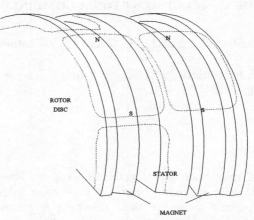

Figure 3   Flux paths in the 'Torus' machine

• **Dimensions.**   A Torus machine design is characterised by the stator core outside and inside diameters, the winding conductor size and number of turns, the pole number, the magnet thickness and material type.

The remaining dimensions and parameters can be determined given values for these quantities. The number of layers of turns needed to form each coil has an important influence on the ease of the winding the stator. The cost of the machine includes material and manufacturing costs. The principal material cost is attributable to the magnets themselves if Nd-Fe-B is used, otherwise the material costs are a relatively small part of the total machine cost.

Losses and Efficiency. The efficiency is determined by the losses, the principal components being:
• I^2R loss in the stator winding
• Eddy current loss in the airgap-wound stator conductors
• Iron loss in the stator core
• Friction and windage

The loss at low and intermediate power is important because the wind turbine spends only a small proportion of its working life at full power. In the present system, however, reduced-speed operation at low power reduces the iron and other losses which are independent of load current thereby greatly improving the part-load efficiency.

The system permits variable-speed operation with shaft power proportional to speed[3]. The generated voltage is proportional to speed and so the current is proportional to speed[2] and the I^2R loss is proportional to speed[4]. At low speed the I^2R loss is less important therefore.

Eddy current loss varies as speed[2], flux density[2] and wire diameter[2]. At low power, the eddy current loss becomes increasingly important, but in the present design it is very low because thin wire is used.

Iron loss varies approximately in proportion to speed.

Rectifier loss is proportional to the dc current and hence to speed[2]

Being a direct-coupled machine, the windage and friction losses are very small.

Losses also determine the winding temperature. But the need for high efficency is a more stringent restriction.

Prototype Design. Many designs were studied using a computer program embodying electric, magnetic and thermal models, the most favourable of those meeting the specification is described by the leading parameters given in Table 1. The magnet thickness was later increased to 12.7 mm to utilise a standard block. Designs using ferrite magnets were sought but no satisfactory solution was obtained within the diameter constraint. However satisfactory solutions do exist for diameters above 650 mm.

Table 1 - Principal dimensions

PARAMETER	VALUE	UNIT
Pole Number	28	-
Winding layers	4	-
Stator Core Outer Dia	465	mm
Stator Core Inner Dia	275	mm
Stator Core thickness	10	mm
Magnet thickness	10	mm

The predicted efficiency over the operating power range is given in Table 2 taking account of the speed and power variation of the loss components as discussed above:

Construction and Testing. The prototype machine is, to the best of the authors' knowledge, the largest of its type yet constructed. It provided valuable experience on the axial forces of attraction which becomes critical in large machines. Ideally, the forces balance but adequate stiffness is needed for stability of the 1·5mm rotor stator clearance. Also, a controlled assembly procedure using jacking screws was needed to resist the attractive force of 10 kN due to the air-gap flux density of 0·69T.

For testing, the generator was driven by a d.c. motor through three stages of speed reduction, using belts for the first two stages and a final chain drive to transmit the high shaft torque of the generator. A torque transducer was fitted on the input shaft to the final drive. The test rig is illustrated in Figure 4

Figure 4 Wind generator on test

Table 2 Efficiency and power over the speed range

Speed rpm	200	180	160	140	120	100	80	60
Shaft power kW	6.41	4.67	3.28	2.20	1.38	0.80	0.41	0.173
Output power kW	5.08	3.73	2.66	1.81	1.14	0.66	0.33	0.129
Efficiency %	79.3	79.8	81.1	82.1	82.6	82.3	80.7	74.5

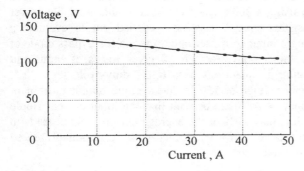

Figure 5 Measured d.c. regulation with resistive load

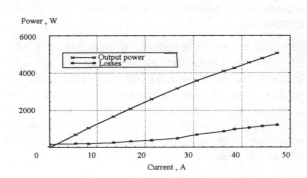

Figure 6 Measured variations of d.c. output power and losses with resistive load

Open-circuit waveshapes, were found to be very close to sinusoidal for line and phase voltages and for a single-turn coil on the stator core. Harmonic analysis of the line voltage waveform showed that the largest harmonic component, the fifth, was only 0·6%. Generator performance was measured under a range of conditions[5]. Figures 5 and 6 give the measured performance at the nominal speed of 200 rev/min with rectifier-resistor load. The d.c. output voltages were 139V on no-load and 107V with rated 5kW load. These were somewhat less than the predicted values owing to variations of some details compared with assumptions made at the design stage. The reduced value of output voltage required increased current (47A) at the rated output power. However, the measured efficiency was 80·7% at this point, exceeding the design prediction of 79·4% even though the latter figure did not include losses in the rectifier and chain drive. This improvement was attributed to the observed sinusoidal waveshape of induced emf which produced a winding current with better form factor.

Power Conversion and Storage

Design of DC Power Conditioner. The dc power conditioner, Fig. 1, uses two step-up converters with their outputs in series. Bypass diodes conduct continuously when the corresponding input source is not available. The design is constrained by stresses on the components due to switching of the two main devices. Analysis of modes of operation[7,8] shows that the worst case occurs when only one of the two sources is operating and the battery supplies the load at nominal voltage. This leads to the highest peak current in both the inductors and the power semiconductors and results in the highest rms current ripple in the output capacitors. These values of voltage and current are used for the selection of power switches and output capacitors, the design of inductors and the prediction of power loss needed for the design of the cooling system.

Selection of power semiconductors. Either MOSFETs or IGBTs would be suitable at this power rating and their costs and drive requirements are similar. IGBTs were selected for their lower voltage drop. A switching frequency of 15 kHz was adopted as a compromise between IGBT power dissipation and size of passive components.

Design of ferrite-core inductors. Inductors rated 850 μH, 40 A are needed for the input circuits and were designed and built to suit. Gapped ferrite cores are usual in such applications to minimise inductor size and losses. The core is assembled from eight ferrite E cores and an adjustable gap is provided to trim the inductance to the desired value. The winding has 44 turns in 4 layers.

Selection of output capacitors. Electrolytic capacitors were selected to provide the required capacitance and voltage and rms current ratings. To meet the target for acceptable heating due to the capacitor ESR loss, each step-up converter stage of the 10 kW prototype uses 3 parallel capacitors each rated 2200 μF - 450 V.

Cooling system design. Design of the cooling system is based on the predicted converter power loss. Assuming rated input of 150V, 33A from each generating unit, it was calculated that the input converters have about 97% efficiency. This drops to about 95% with one unit idle because of increased IGBT conduction loss at the higher duty cycle.

For natural convection, the power switches are to be mounted on an aluminium heatsink with thermal resistance of 0.15 ºC/W to air. Since the inductors and capacitors are housed inside the converter frame, internal air circulation is important and is created by two low-power fans controlled by a temperature sensor.

Construction and Testing. Construction of the 10 kW prototype has been completed and laboratory tests have been were carried out to assess performance before assembling the complete system for field tests. Fig. 7 shows the prototype converter: The output inverter IGBT modules and rectifiers mounted on the aluminium heatsink can be seen in the upper part of the figure The input converters auxiliary circuitry can be seen in the lower part.

Fig. 7   Prototype 10kW power conditioner.

A laboratory test rig was built using variable transformer-rectifier power supplies to emulate the PV array and the wind generator. A dc load was arranged by using a 300V lead-acid battery and available power resistors. Initial tests were carried out to evaluate a single input converter operating via the bypass diode of the other converter. Fig. 8 gives traces of input and output dc quantities with IGBT duty-cycle set at 0.5, input voltage of 150 V and average output current of about 14 A. It can be seen that the input current ripple is less than 20% of the average current, which leads to acceptable converter losses with a reasonably sized input inductor.

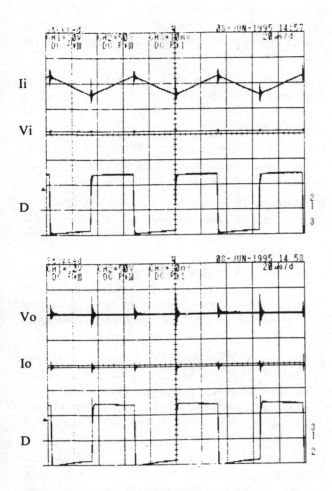

Fig. 8   input and output dc quantities
time scale 20 µs/div,   voltage 50 V/div:
current 5 A/div,   D = IGBT signal

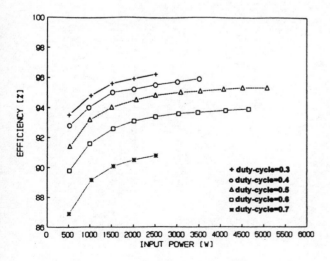

Fig. 9   Curves of efficiency vs. input power of a single
dc-dc converter.

Performance was also evaluated with both input circuits active. These tests were carried out using the actual

values of input voltage and current resulting from the design and construction of the two generating units. The reduced value of the input voltage at which both the wind and PV units delivers the 5kW rated power (i.e. 107V and 136V, respectively) required increased input currents, and leading to reduced efficiency in the converter. Fig. 10 shows efficiency curves achieved from tests in which a dummy PV unit delivers constant input power (i.e. either 1kW, or 3kW or 5kW) while the wind unit produces variable power output. It can be seen that the efficiency achieved by the prototype unit remains between 94% and 95% over a wide power range.

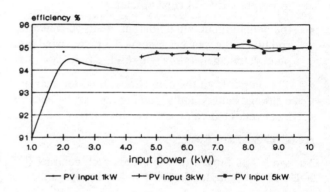

Fig. 10 Curves of efficiency vs. input power of the 10kW power conditioner prototype.

Single converter tests were conducted across a range of input powers from 500W to 5kW. Efficiency, calculated from the measured input and output powers is plotted in Fig. 9. It is found that with 5 kW input power and a duty-cycle of 0.5 for the IGBT each dc-dc converter has efficiency slightly better than 95%, which is in good agreement with the predicted value.

Battery Selection. A 150Ah, 300 V storage battery has been selected for a future field test programme. The voltage is chosen to cause the input converters to operate with acceptable duty cycle and the capacity is chosen with a view to envisaged applications of such combined wind/PV generating systems.

Output Inverter. The output inverter is required to produce a sinusoidal output voltage waveform with fixed both amplitude and frequency. Single-phase or three-phase applications will arise. Suitable sinusoidal PWM inverters are commonly used for uninterruptable power supplies therefore no development is needed in this area. The project has therefore concentrated on the provision of the dc power and its delivery to the main dc link. Testing has been carried out simply using dc loads.

Although the proposal and the 10 kW pilot plant relate to stand-alone applications in which the wind generator and PV array are the available sources, it would be possible to combine the plant with other sources by connection at the dc link or by connection at the ac terminals of the inverter. Possibilities include diesel, bio-fuel or mini-hydro plant. The energy storage requirements would of course be altered.

4. CONCLUSIONS

The design and construction and testing of the prototype components of a 10 kW pilot plant has been described. Unconventional solutions adopted for the wind generator and the power conditioner have been outlined. The principal features of the proposed plant are:

- the wind generator is a directly-driven, multipole, permanent-magnet machine which has high torque to weight ratio and good efficiency.

- the double-input, single-output, dc-to-dc converter, combines the power generated by the wind and photovoltaic generators with high-efficiency

- The output from the converter is used for charging a storage battery and supplying the user ac load via a conventional voltage source inverter;

ACKNOWLEDGEMENTS

The work was funded through a research contract (CT JOU2 93-0420) under the EU JOULE II programme.

REFERENCES

1. Spooner, E.; Williamson, A.C.; Thomson, L.: "Direct-Drive, Grid-connected, Modular Permanent-Magnet Generators", British Wind Energy Association Conference, Stirling, June, 1994.

2. Spooner, E.; Williamson, A. C.: "Direct-coupled, permanent-magnet generators for wind turbine applications", Accepted for IEE Proc-B Electric Power Applications.

3. Spooner, E.; Williamson, A. C.: "Modular design of permanent-magnet generators for wind turbines", Accepted for IEE Proc-B Electric Power Applications.

4. Spooner, E., Chalmers, B.J.: "'TORUS': A slotless, toroidal-stator, permanent-magnet generator", IEE proc B, 139, 6, Nov 1992, pp497-506.

5. Chalmers, B. J.; Wu, W.; and Spooner, E.: 1996: "An axial-flux permanent-magnet generator for a gearless wind energy system". PEDES '96, New Delhi, India.

6. Di Napoli, A.;Caricchi, F.; Crescimbini, F.; Noia, G.: "Design criteria of a low-speed axial-flux PM synchronous machine", Proc. Int Conf Evolution and modern aspects of synchronous machines SM100 Zurich Aug 25-27, 1991, pp 834-839.

7. Caricchi, F. Crescimbini, O. Honorati, E. Santini "Design and testing of a small-size wind-photovoltaic system prototype", Proceedings of the European Community Wind Energy Conference and Exhibition 1993, Lubeck-Travemunde (Germany), 8 - 13 March 1993, pp. 740-743.

8. Caricchi, F. Crescimbini, A. Di Napoli, O. Honorati, E. Santini "Testing of a new dc-dc converter topology for integrated wind-photovoltaic generating systems", Proceedings of the Fifth European Conference on Power Electronics and Applications, Brighton (UK), 13-16 September 1993, Vol. 8, pp. 83-88.

PERFORMANCE ASSESSMENT OF VARIABLE SPEED WIND TURBINES

B Connor and W E Leithead

University of Strathclyde, UK

Abstract

Renewable sources of electricity generation, of which the most promising for the UK is wind energy, are becoming increasingly important due to concern over the environment. The purpose of this paper is to assess the performance of variable speed wind turbines for several operational strategies. Two contrasting wind turbine rotors are considered. The performance is assessed by simulation.

1 INTRODUCTION

Due to the environmental problems associated with fossil and nuclear fuels, there has been an increase in interest in renewable sources of energy. Wind energy has been the most successful of all renewable sources, especially in the UK, where 40 % of Europe's wind resource can be found. The installed capacity in Europe has increased considerably and is currently at 1400 MW [1] with the UK contributing about 140 MW [2]. Twenty two wind farms consisting of 400 wind turbines are currently in operation or under construction in the UK. The farms have the potential to generate 360 GWh annually and to satisfy the need of approximately 250,000 consumers.

The wind resource is of course free and non-depleting and does not produce any harmful products. The effect of the UK wind energy programme is to reduce the CO_2 emissions by 400,000 tonnes annually. The projected installed capacity within the UK by the year 2005 is approximately 20 TWh/year with the cost of electricity about 4p/KWh [2] which is comparable with coal and nuclear generated electricity. The future for wind energy therefore remains bright. The main stumbling block in the development may be in the public perception of wind turbines. The main objections by the public are the extent of noise and visual pollution. However, these obstacles can be overcome to a certain extent by building wind turbines offshore, where the wind speeds are considerably larger, or by employing variable speed operation.

2 VARIABLE SPEED WIND TURBINES

There are essentially two types of wind turbines - a constant speed wind turbine and a variable speed wind turbine. In the case of a constant speed wind turbine, the rotor is connected, by way of a gearbox and a generator, directly to the electrical grid. As a consequence, the rotational speed of the wind turbine is fixed by the grid frequency. In the case of variable speed wind turbines, the generator is connected indirectly to the grid via a DC-AC link. Hence the rotor speed is independent of the grid frequency. In this paper, the performance of a three bladed, medium sized, horizontal axis grid connected wind turbine with variable speed operation is assessed.

The ability to vary the rotor speed increases the operating flexibility of the wind turbine and offers several advantages over constant speed machines. Firstly, the energy available in the wind depends on the tip-speed ratio, λ , where $\lambda = \frac{VR}{U}$. R, V and U represent the rotor radius, rotor speed and wind speed respectively. Variable speed wind turbines can be controlled to regulate the rotor speed such that λ attains its optimal value and hence optimises energy capture. The ability to vary the rotor speed also means that the audible noise levels are reduced since these are dependent on wind speed. The effect of noise pollution is greater in light wind conditions. The advantage of variable speed wind turbines in high wind conditions is in the ability to alleviate loads and hence increase the life time of the wind turbine. However, these benefits are incurred at the expense of some disadvantages. Namely, the variable speed drive is expensive, although it is anticipated that the price will fall as technology advances. Also, due to the indirect grid connection, the drive-train dynamics are lightly damped which may result in mechanical vibration. There is a great variety of wind turbine configurations. For example, as well as being constant speed or variable speed, the wind turbine might be full span pitch regulated, tip regulated or stall regulated; three-bladed or two-bladed. The dynamics and control of each configuration are different. This paper is concerned with the performance of three bladed variable speed wind turbines both as a full span pitch regulated machine and a stall regulated machine. The machines have the same power rating of *300 kW* .

3 WIND TURBINE OPERATIONAL MODES

The aim of the control system in light winds is the same for both pitch regulated and stall regulated wind turbines, namely, to increase energy capture. The power available in the wind, P, is related to rotor speed, V, and wind speed U, by the relation

$$P = \frac{1}{2}\rho\pi R^2 U^3 C_p(\lambda)$$

where ρ is the air density. C_p is the power coefficient and is a function of the tip-speed ratio. The power from the wind is maximised if the rotor speed is such that C_p is maximised. The shape of the C_p-λ curve has a significant

Opportunities and Advances in International Power Generation, 18–20th March 1996,
Conference Publication No. 419, © IEE, 1996

influence on energy capture [3]. In the case of a pitch regulated rotor, the C_p-λ curve is designed to have a broad flat peak, whereas the C_p-λ curve for a typical stall regulated rotor has a sharp peak. This suggests that the efficiency of a stall regulated wind turbine is sensitive to small deviations from peak C_p. The below rated control system is a single-input single-output problem.

In the case of pitch regulated wind turbines, the variable speed capability is exploited above rated wind speed to minimise the drive-train load transients whilst the variable pitch capability is exploited to restrict the speed of the rotor. The control system is designed to maintain the rotor speed to within a small band of its rated value, typically 5%. The associated control system is multivariable with two inputs and two outputs.

In above rated operation, variable speed may be employed to stall the wind turbine whilst keeping the pitch fixed. There are several strategies which can be adopted to stall the wind turbine [5], however, this paper considers only two. The first is appropriate for a rotor having a sharp C_p-λ curve. The overall strategy is to maximise energy capture until the rotor speed reaches $3.5\ rad/s$. The rotor speed is then actively controlled in high wind speed such that the generated power is less than its rated value. The reason for this is that the dynamics allow for better overall controller performance as compared to maintaining power at $300\ kW$ [5]. The second strategy is adopted for a rotor having a flat C_p-λ. The aerodynamics of this rotor are such that the strategy adopted by the sharp C_p-λ curve is not suitable.

The reason is that high torque transients are produced as the rotor speed is reduced to stall the wind turbine. It is therefore required to adopt a strategy which avoids these large torque excursions. The most appropriate strategy for this rotor is to optimise energy capture in low wind speeds until a specified rotor speed is reached. The rotor speed is held constant at this value until rated power is reached. The rotor speed is actively controlled in high winds such that the power is maintained at rated. Note that maintaining the power at its rated value in this case produces adequate results and therefore it is not necessary to adopt a strategy which generates less power.

The operational modes described above are assessed thoroughly in the following section by simulating with several realistic wind regimes in order to express the performance in terms of wind speed.

4 PERFORMANCE ASSESSMENT

Variable speed wind turbines are perceived to have two main potential advantages: increased energy capture in low wind speeds and the reduction of drive-train loads in high wind speeds. A more precise statement of the potential advantages is required. Consider the first potential advantage of increased energy capture in below rated wind speed. The absolute level of energy capture is not particularly apt since changes to the design of any wind turbine which increases its energy capture are readily envisaged, e.g. increasing the rotor diameter. Rather, it is whether increased aerodynamic efficiency enables a more

cost effective design of wind turbine to be realised. Similarly, for the second potential advantage of reduced drive-train loads in above rated wind speeds, it is whether decreased mean drive-train load perturbations enable a more cost effective design of wind turbine to be realised. The mean loads are, of course, essentially the design loads. It follows that appropriate measures of performance for variable speed wind turbines are

- aerodynamic efficiency in below rated wind speed
- extent of transient drive-train loads in above rated wind speed

The performance of a particular strategy is critically dependent on the rotor characteristics and on the wind regime driving the wind turbine. Having determined the above two measures, the most appropriate operational mode of a variable speed wind turbine is assessed as follows [4]. The performance of the wind turbine in the operating modes described in Section 3 is assessed by a full non-linear simulation. The simulation runs consist of differing wind regimes, namely with mean wind speeds from $5\ m/s$ to $23\ m/s$ in steps of $2\ m/s$. The data is sorted in ascending order of wind speed and divided into bins of wind speed of $0.5\ m/s$. Various statistics are computed in each bin. The statistics presented in this paper are the mean efficiency and the mean low-speed shaft torque. This process is repeated for several turbulence levels [4]. In this paper, the results corresponding to a turbulence intensity of 20% are presented which represents harsh wind conditions.

The efficiency, defined in this paper, has two definitions - one for below rated wind speed and the other for above rated wind speed. In the following, the aerodynamic efficiency, η, is defined below rated as the ratio, expressed as a percentage, of energy capture to the energy capture that would be attained were C_p always to have its maximum value; that is $\eta = \overline{C_p}/C_{p\,max}$ where $\overline{C_p}$ is the mean power coefficient. In above rated operation, efficiency has a different definition since it is not required to extract all the power from the wind. In this case, the efficiency is defined as the ratio of the mean power attained to the rated power of the wind turbine i.e. $\eta = \overline{P}/P_r$ where $\overline{P}$ is the mean power and P_r is the rated power, namely $300\ kW$.

4.1 Results of pitch regulation

The performance of the pitch regulated rotor is described by Fig. 1(a) and Fig. 1(b). Each point represents the mean value in each bin. For example, the first point corresponds to a wind speed of $4.25\ m/s$ and the last corresponds to $33.25\ m/s$. The rated wind speed i.e. the wind speed at which the power in the wind is $300\ kW$, occurs at $11.5\ m/s$. Fig. 1(a) shows a drop in efficiency between $11\ m/s$ and $12\ m/s$. The reason is that the control system switches from below to above rated operation and the efficiency is reduced as a result. The aerodynamic efficiency of a wind turbine is assessed by the efficiency for wind speeds less than rated wind speed. For the pitch regulated rotor considered in Fig.

1, η varies from *98.0 %* to *99.6 %* . The high aerodynamic efficiency of the pitch regulated wind turbine is expected since the rotor has a broad flat C_p-λ curve indicating that the efficiency is insensitive to small deviations from $C_{p\ max}$. The aim of the pitch regulated wind turbine in wind speeds greater than *11.5 m/s* is to achieve good power control and also adequate drive-train torques. The efficiency above rated varies from *90.2 %* to *108.4 %* which means that the generated power varies from *270.6 kW* to *325.2 kW* . The drive-train torque is represented by the low-speed shaft torque as shown in Fig. 1(b). The mean drive-train torque obtained in above rated operation compares well with the rated drive-train torque of *65.2 kNm* .

4.2 Results of stall regulation

The performance of the stall regulated wind turbine with sharp C_p-λ is shown in Fig. 2(a) and Fig. 2(b). The rotor speed is reduced in wind speeds greater than *11.2 m/s* in order to generate less than rated for the reason given in section 3. It is shown in Fig. 2(a) that the aerodynamic efficiency below rated varies from *93.0 %* to *97.8 %*. As expected, the efficiency is less than that of a rotor with a flat C_p-λ curve. The power overshoots just after the switching point and is shown in Fig. 2(a) for wind speeds in the range *11 m/s* to *19 m/s* . The efficiency above rated varies from *76.9 %* to *109.5 %* which corresponds to a range of generated power of *230.7 kW* to *328.5 kW* . The performance measure of adequate drive-train levels in above rated wind speeds is assessed in Fig. 2(b). The drive-train torque varies from *55.0 kNm* to *77.1 kNm* . Fig. 2(a) and Fig. 2(b) show that a rotor having a sharp C_p-λ curve performs well when operating as a variable speed stall regulated machine although the aerodynamic efficiency is reduced. As a contrast, the performance of stall regulated wind turbine with a flat C_p-λ curve is assessed in Fig. 3(a) and Fig. 3(b). The rotor is the same as the pitch regulated rotor and therefore the below rated efficiency is the same. The strategy adopted here is to track the $C_{p\ max}$ curve until the rotor speed reaches *3.4 rad/s* i.e. the wind speed is approximately *9 m/s*. The rotor speed is held at this value until rated power is reached at a wind speed of about *15 m/s* . In high wind speeds the rotor speed is reduced in order to stall the wind turbine. The effect of the constant speed mode on efficiency is to reduce it since efficiency is not being maximised over all below rated wind speeds. The efficiency varies from *88.4 %* to *156.1 %* in above rated wind speed. The corresponding generated power varies from *265.2 kW* to *468.3 kW* . The large peak power is due to the wind turbine being subjected to large off-design aerodynamic torques and is a characteristic of the rotor aerodynamics. The poor performance is also apparent in the drive-train torque histogram shown in Fig. 3(b) which shows a variation of torque between *84.2 kNm* and *171.7 kNm* . Fig. 3(a) and Fig. 3(b) show that a rotor having a broad flat C_p-λ curve does not perform well when operating as a variable speed stall regulated wind turbine.

5 CONCLUSIONS

Variable speed wind turbines have the potential advantages of increased efficiency and acceptable drive-train loads. In order that wind energy can compete with fossil fuels in terms of price, the wind turbine energy conversion system must realise these advantages to within a desired tolerance. In this paper, the performance of a pitch regulated and stall regulated wind turbine is assessed by expressing the efficiency and drive-train torque as a function of wind speed. Assessing a wind turbine in this way allows for a particular wind turbine and operational mode to be compared and the most suitable shown. In particular, if the wind distribution of a given site is known, then the efficiency can be used to predict the energy yield per annum of the wind turbine. The paper has shown that a rotor having a flat C_p-λ curve has high aerodynamic efficiency and is suited for operation as a pitch regulated machine. A rotor of this type is not suitable for operation as a stall regulated wind turbine. A rotor having a sharp C_p-λ curve is more suited to operation as a stall regulated wind turbine although the aerodynamic efficiency is reduced.

ACKNOWLEDGEMENTS

The DTI (formerly D.En) and ETSU, by whose permission this paper is published, are gratefully acknowledged for supporting the work presented in this paper

REFERENCES

[1] Diamantaras, K., *Wind energy in the European market*, Proc., 16th British Wind Energy Association Conference BWEA16, Stirling, 15-17 June, pp. 9-12, 1994

[2] Lindley, D., *Wind energy - current status in the UK*, Proc., 16th British Wind Energy Association Conference BWEA16, Stirling, 15-17 June, pp. 1-8, 1994

[3] Connor, B., Leithead, W.E., *The effect of rotor characteristics on the control of pitch regulated variable speed wind turbines*, Proc., 16th British Wind Energy Association Conference BWEA16, Stirling, 15-17 June, 1994, pp. 67-72

[4] Connor, B., Leithead, W.E., *Strategies for the control of variable speed HAWTs*, Report prepared for the DTI (ETSU) under contract No. E/5A/6097/2750, Feb. 1995

[5] Connor, B., Leithead, W.E., *Control strategies for variable speed stall regulated wind turbines*, Proc., European Wind Energy Association Conference EWEC'94, Thessaloniki, Greece, 10-14 October, 1994

68

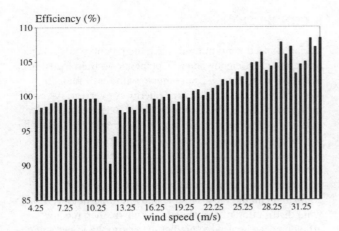

Fig. 1(a) Efficiency histogram for pitch regulated wind turbine (flat C_p-λ curve)

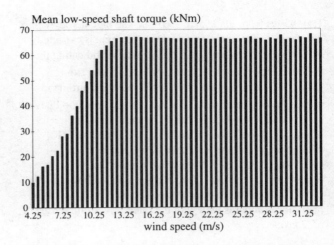

Fig. 1(b) Drive-train torque histogram for pitch regulated wind turbine (flat C_p-λ curve)

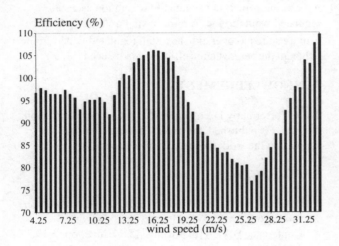

Fig. 2(a) Efficiency histogram for stall regulated wind turbine (sharp C_p-λ curve)

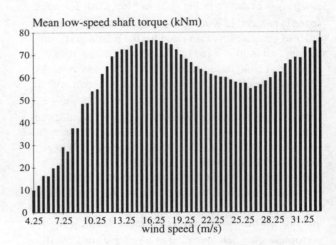

Fig. 2(b) Drive-train torque histogram for stall regulated wind turbine (sharp C_p-λ curve)

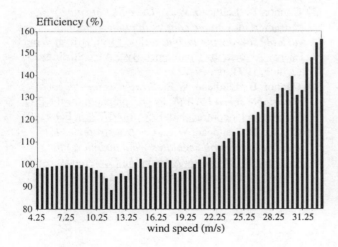

Fig. 3(a) Efficiency histogram for stall regulated wind turbine (flat C_p-λ curve)

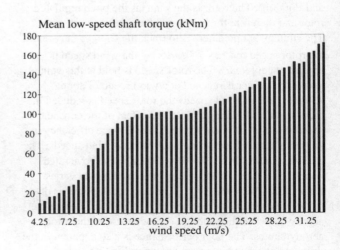

Fig. 3(b) Drive-train torque histogram for stall regulated wind turbine (flat C_p-λ curve)

EXPERT SYSTEM SCHEDULING OF CASCADE HYDRO-ELECTRIC PLANTS

M W Renton, A R Wallace

Napier University, University of Edinburgh

The hydrological interdependence of plants in cascade hydro electric systems means that operation of any one plant has an effect on water levels and storage at other plants in the system. Hydro plants may be dispatched to meet predicted demand, to trade electricity at a commercially opportune time, or to adjust water levels in the system. Hydrologically and commercially efficient operation of cascade systems requires that water resources and energy trading are managed simultaneously. This paper describes some of the principles of cascade water management and the development of an expert system which assists in the optimal scheduling of hydro plant within Scottish Hydro Electric plc's system.

ENERGY TRADING MANAGEMENT

Up to 25% of the annual energy production by Scottish Hydro Electric (SHE) can be from conventional hydro-electricity generation, provided by 33 hydro systems mostly within 8 cascade schemes, consisting of 76 reservoirs and 54 power stations with a total installed capacity of 1025MW. Control of these hydro plants is undertaken at two production group control centres at Dingwall and Clunie. The Dingwall Hydro Group controls the northern area incorporating the Shin, Conon, Affric-Beauly, Garry-Moriston and Foyers schemes while the Clunie Hydro Group controls the southern schemes of Tummel, Breadalbane and Sloy-Awe, [1]. Electricity demand varies continually and must be met at all times by the combined output of hydro, gas, coal, oil or nuclear plant. Complex contractual constraints set minimum levels of gas-fired and nuclear generation which are not intended to vary on a day-to-day basis and these meet a portion of overall demand. Engineers in SHE Central Control Room in Pitlochry call for generation from all hydro plants and those operating on other fossil fuels to meet total demand. Traditionally hydro plant has been used as high-intermediate load capacity, since the fast access times commend its application to meeting short term demands with economic efficiency. The availability of an energy source with this flexibility allows SHE to meet variations in demand largely from hydro resources.

In association with the Hydro Groups a Dispatch Schedule is formed which takes account of any system limitations or hydro plant restrictions. The Groups allocate generating limits to their individual power

stations in the form of a generation profile. The required output in MW from each power station is listed for every half-hourly increment over a week and each plant has 336 slots of scheduled generation allocation.

Thus the hydro generation programme may be altered at any time during the day as the stations are dispatched or shut down to meet anticipated increases or reductions in demand. In addition at times of peak demand, the pool price for electricity is high reflecting the RECs need to purchase energy. At these times there is opportunity to dispatch additional hydro plant with commercial advantage. During troughs in demand the price is correspondingly lower and there is opportunity to reschedule plant and conserve water. The day is divided into eight time periods covering these peaks, troughs and periods of intermediate demand. The start time and length of these periods can vary and the shape of the demand curve changes for different seasons and days of the week. Thus, even if the Dispatch Schedule has been produced, it regularly has to be changed at short notice. Optimum scheduling of plant relies heavily on the experience and estimating capabilities of key personnel to quickly determine the effect on the water system should the status of any hydro plant be changed.

WATER RESOURCE MANAGEMENT

Water management of an individual reservoir and hydro plant is a well established practice, where the water throughput of the station, together with historical hydrological data are used to calculate dynamic changes in level, energy storage and volume of the reservoir. However, the combination of multiple reservoirs, rivers, weirs and plants in a cascade becomes more complex since the level of each reservoir, or plant output, relies heavily on the dynamics of upper and lower systems. Control engineers rely heavily on their experience and understanding of the hydraulic dynamics of individual and interconnected reservoirs. Another important function of the Hydro Groups is the management of stored energy in the SHE reservoirs. Optimisation of the revenue from hydro generation requires that energy stored as water does not become wasted by spillage or fail to attract its highest market value by generating unnecessarily when demand and prices are low. Furthermore since the reservoir water levels and river flows have an impact on the natural environment SHE must ensure that dispatch of a hydro plant does not

Opportunities and Advances in International Power Generation, 18–20th March 1996,
Conference Publication No. 419, © IEE, 1996

suddenly raise or lower water levels, or allow them to move outwith agreed limits. Correct scheduling of hydro plant requires the accumulated operating experience of the engineers who refer to large amounts of data describing the current status of the system together with records describing the history of operation. There is also a higher level of knowledge that the engineers draw on, which is based on their instinct and the ability, to forecast from experience what the likely outcome of a decision or schedule will be. With training and experience the operating skills are slowly transferable, but the ability to be able to forecast or practice insight is a highly personal skill which takes considerably longer to build up. Expert systems have been applied successfully to other system control and managerial processes to support the decisions as they are made and to allow the testing of prospective system settings. In addition such systems allow the recording and transfer of experience and skills to other staff.

A Water Manager (WM) Decision Support System has been developed with SHE which contains system detail and operator-based knowledge, and can assist control engineers to better utilise hydro resources by:
1) determining the most advantageous operating schedule,
2) using and maintaining on-line up-to-date operational information,
3) taking account of all hydrological factors and variables,
4) ensuring all environmental constraints are met,
5) calculating and testing the best course of action to take, when a change in operation occurs

RESERVOIR DYNAMICS

Naturally occurring reservoirs are not standard geometric shapes. Generation and associated discharge varies with time and there are environmental and weather factors which also contribute directly to the change in water level. A reservoir thus has a number of independent inflows and outflows as shown in figure 1. In the case of SHE reservoirs, these operational variables are:
(i) Inflow from upper power stations (controlled by SHE)
(ii) Outflow from lower power stations (controlled by SHE)
(iii) Runoff from the surrounding catchment (weather dependent)
(iv) Rainfall and Evaporation (weather dependent)
(v) Compensation inflow/outflow (decreed by national or local By-laws)
There are frequently changes in (iii) and (iv) to which SHE must react by adjusting (i) and/or (ii) to maintain the water levels in all reservoirs and rivers, subject to

the prescriptions in (v). The levels in reservoirs and rivers are measurable variables and effective water management seeks to maintain them within prescribed limits. This requires the discharge from upper and lower hydro plants to be predictable and controllable and, taking account of the weather and environmental factors, used to forecast the effect on upper, lower and interlinked reservoirs.

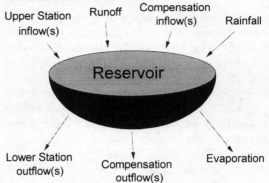

Figure 1 - Reservoir Dynamics

RESERVOIR VARIABLES

If water levels are to be managed effectively the following data is necessary for every reservoir in the system which, in most cases, has been accumulated over a long period of time.

Storage and Volume: The storage, in millions of cubic metres (MCM), must be calculated from the topography of the reservoir bottom. Since the change in volume with level is non-linear, "look-up" tables are used to relate level to stored volume. The head (m) on the turbine(s) and the discharge ($m^3.sec^{-1}$) determine the power developed (kW). Over an elapsed time the energy converted (kWh) is related empirically to the volume of water discharged (m^3). Each reservoir has a discharge/energy characteristic ($m^3.kWh^{-1}$) and thus the stored energy may be estimated for any measured level.

Run-off: There is a Long Term Annual Average Run-off (LTAA) from the land into the reservoir which is a fixed value of kWh, usually calculated by measuring the generated energy which maintains the level over a year, taking account of compensation flow and intentional spillage. A monthly average is calculated from the LTAA run-off which is multiplied by a monthly factor to give a value for the 100% run-off for that month, but this may be adjusted depending on the weather and environmental conditions at any particular time to determine a run-off for the month concerned. Where more extensive records are available a 10-year average runoff is used.

Weather: The local rainfall/evaporation effects on any reservoir are generally unpredictable. Several methods have been devised to model the evaporation from large volumes of water but each tends to be site specific, depending on land relief and ambient conditions, [2]. Since there is sufficient historical run-off data for reservoirs in the SHE area weather effects are implemented by adding or subtracting an appropriate percentage of the total run-off.

Generation Inflow and Outflow: Generation flows are established by multiplying the turbine discharge/energy characteristic ($m^3.kWh^{-1}$) of the reservoir by the time integral of turbine output power (kWh), estimated from the generation profile of the plant.

Compensation Flow: Each reservoir may have a fixed value of compensation inflow or outflow to maintain the appropriate river flow, which may change with season.

Levels: The reservoir level is calculated at the end a given operating time and compared with a number of levels which define the course(s) of action which ensure correct water management. These levels are, [3]:
Maximum or Spill Level at which the water in the reservoir begins to spill over the edges or through a spillway possibly flooding the adjacent area.
Full Generation Level above which the plant should operate at full capacity. Failure to schedule plant accordingly would eventually lead to spillage.
Target Level which makes best use of the run-off. This level is estimated by taking account of all foreseen outages of plant, predicted demand and average monthly run-off.
Maximum and Minimum Normal Levels are the preferred values but these may vary seasonally or for operational reasons.
Minimum Duty Level is the lowest that the reservoir may fall to while retaining enough water for emergency generation and compensation flow.

A further operational constraint on any reservoir is the rate-of-rise or -fall of water level. This is important where a river/reservoir is used by the public. While a gradual change in level goes unnoticed, a fast change can cause problems such as grounding of boats or inundation.

CASCADE HYDRO SYSTEMS

The reservoir dynamics model in the WM was originally developed using the Quoich-Garry cascade system in the Garry/Moriston Scheme, shown in figure 2. Modelling cascade serial schemes with a combination of multiple reservoirs, rivers and hydro plants relies heavily on reservoir dynamics of upper and lower systems. This is complicated further by the reservoir time constant, which is defined as the time taken for inflow entering the reservoir to be experienced at the outflow. Similarly there is a longer time constant relating the time taken

for rainfall over a catchment area to run off the land and produce a rise in level. Finally, since water flows from higher reservoirs down through the system to the lower reservoirs, the tendency might be to analyse the upper reservoir dynamics then assess the effect on the next lowest reservoir, repeating this through the system. In more complex schemes, upper reservoirs may be in isolated areas and lower reservoirs within populated or recreational areas. Rapid or extreme level changes may be intolerable. Thus the highest priority reservoir may be in the middle of the cascade while the lowest priority is at the highest point. Analysis of the dynamics is done in order of priority, to first ensure stable operation of the most important reservoirs. In the long term this has no bearing on the flows, but in crisis water management it has a significant effect on transient flows and levels.

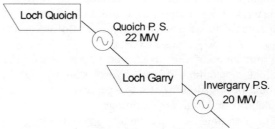

Figure 2 - Serial Cascade: Quoich/Garry Cascade

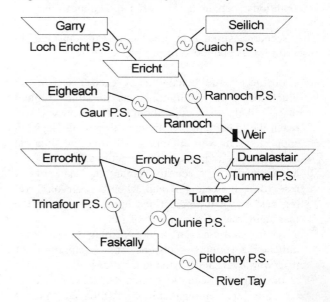

Figure 3 - Tummel Valley Scheme

Since most of the SHE Schemes do not form simple serial cascades, the WM Decision Support System can represent any scheme with multiple reservoirs, hydro plants and weirs in series/parallel combination. Figure 3 shows one such scheme.

The model of such a scheme determines the change in all reservoir water levels taking account of availability and limitations on generation, variable weather and environmental conditions, the interrelation between hydraulically linked reservoirs and reservoir priority.

THE WATER MANAGER

Using Object-Oriented programming language and a forward chaining rule-base system, SHE operational expertise and decision-based logic has been combined into a PC based decision support system which replicates the hydrology and decision processes involved in water management and hydro plant dispatch. The WM has the following presentation and operational features:

❶ a user-friendly graphical interface.
❷ an on-line help system.
❸ ability to install (or delete) and model any cascaded hydro scheme .
❹ dialogue boxes within which the user can view and update reservoir, station and weir data attributes.
❺ easy access to associated application software, i.e. Lotus 1-2-3 and Microsoft Write.
❻ determination of the best generation profile by optimal scheduling of all hydro stations, taking account of priority time slots, target levels, station set availability and limiting conditions
❼ scenario testing of operation for any scheme given the: start-time, duration, incremental time, weather conditions, percentage runoff, and data location.

The scenario simulation determines the change in reservoir levels and the WM provides:

① a profile of the behaviour of each reservoir, in graphical form, highlighting any possible problem that is likely to occur. Should any reservoir level reach the maximum or minimum limits, the decision support system will attempt to alleviate the situation and report the required avertive actions.
② an annunciator screen to display the state of all reservoirs, indicating if and when a reservoir level reaches any of the prescribed levels or exceeds the maximum rate-of-rise or rate-of-fall.
③ a record and on screen log display of all level indicators, support decisions and a summary of reservoir levels and station generation.
④ at the end of a scenario simulation, the contents of the log are automatically downloaded into a numbered/dated text file.

The decision support software includes many rules which reason towards the goal of preventing a reservoir from spilling or draining. For example, if a reservoir was about to spill there are five possible actions:

     Reduce generation from an upper plant.
     Increase generation at a lower plant.
     Raise an upper weir.
     Drop a lower weir.
     No change
The fifth option arises as some plants must operate at the same output throughout the day to prevent changing the noise levels or altering the natural flow of a river

thereby creating an environmental disturbance. The consequences of a change to the operating plan will obviously affect the dynamics of the connecting upper or lower reservoir(s). Reservoirs are prioritised to the extent that if a reservoir is about to spill to save another adjoining reservoir, then the lower priority reservoir would be left to accept the consequences. Although this situation rarely arises, in extreme cases the option has to be considered. Thus during a scenario simulation the decision support system will indicate and record the appropriate action to take to prevent spillage or draining of reservoirs. It will also automatically change the generation profile if appropriate. These changes can then be reincorporated into an operational plan and the simulation can be rerun to ensure better viability.

CONCLUSIONS

Hydro plant is now being operated more flexibly than in the past to meet commercial, technical and hydrological requirements. This may result in a varying priority of dispatch which frequently bears little resemblance to the normal roles and positions in the traditional merit order.

The consequences of a new dispatch schedule on the hydrological distribution of water in the cascade system must be predicted rapidly and consistently, taking account of initial storage levels and weather effects which vary seasonally.

It is possible to incorporate system data and operational practice into a software model which will forecast the response of the system with acceptable accuracy, repeatability and confidence.

ACKNOWLEDGEMENTS

The authors would like to thank Scottish Hydro-Electric plc for their financial support and technical assistance.

REFERENCES

1. NSHEB, "Power From The Glens", North of Scotland Hydro-Electric Board Publication, 1990.
2. Takhar, H.S and Cook, D.J.; "Investigation of Some Evaporation Models in the Hodder Catchment", Israel Technical Journal, volume 22, 1984/85.
3. Johnson, F.G. and Cooke, W., "Operation of the Reservoirs of the North of Scotland Hydro-Electric Board"

ENERGY FROM WASTE IN THE SEWAGE TREATMENT PROCESS

S. WALKER

Thames Water, England

INTRODUCTION

Established in 1974 and transferred to the private sector in 1979, Thames Water is the provider of water services to seven million and sewerage services to over eleven million customers in London and the home counties. These services are provided through a network of 30,000 kms of water mains and 83,000 kms of sewers, stretching from east London to Cirencester in the west, Banbury in the north and Crawley in the south. The energy source is methane, which is produced from the anaerobic digestion of sewage sludge at 300 treatment works throughout the Thames Water region.

The roots of the company's interest in energy are easy to understand. Thames Water consumes over £40m of electricity per year. 70% of this demand is expended on pumping alone, while much of the remainder is needed to aerate sewage as part of the treatment process. The power generated by the waste is worth a further £8m per annum to Thames Water. Naturally the efficient and economic use of energy is high on the company's agenda.

Thames Water and its predecessors have been generating electricity from sewage for half a century. At Mogden Sewage Works near Twickenham, London, dual fuel engines were installed as long ago as 1936 and remained in operation until they were replaced by modern turbo-charged units less than ten years ago. Those engines were the first of their kind in Europe. The original engines operated successfully for more than 40 years providing electricity for site use and waste heat to keep the anaerobic digesters at the required temperature.

POWER GENERATION PROCESS AND PLANT

In the process of treating sewage, water is separated from the incoming effluent in large sediment tanks. The sludge produced in this way is then rendered harmless by the next stage of the process in the digestion plant. In the digesters, where sludge is maintained at a temperature of 25-28°C for some 48 days, bacteria break down the organic matter to produce a mixture of methane and carbon dioxide.

It is this bio-gas which is compressed and used to

fuel generating plant, with the waste heat from the engines being used in turn to keep the digesters warm. Any excess heat is used by the site's heating system. Power generation plants are now installed at 23 of the largest treatment works in the Thames Water area, with sludge from some of the smaller works being piped or tankered to the larger sites to enhance gas production.

The type of engines installed, to a large extent, is a factor of size of the site and amount of sludge available. At the largest site, Beckton, which serves a population of 2.8m, we have found it feasible to install industrial gas turbines with steam turbines running in combined cycle mode, producing 6-7 MW. At the other end of the scale, for example at Bracknell, we have found it appropriate to install spark ignition packaged plant producing some 100 kW. Apart from the size, the choice of plant is dictated by the bio-gas composition and in particular the H_2S content which can have a devastating effect on engine life. Engines most tolerant of the impurities in the fuel are the dual fuel compression ignition engines which are predominantly used in the mid- to upper-sized ranges.

As the largest operator of sewage gas generating plant Thames Water has amassed an unparalleled operating experience, which is quite often drawn upon by equipment manufacturers. In recent years - particularly with the introduction of non-fossil fuel obligation (NFFO) - a greater focus on the power generation process and better management has enabled a 25% increase in output to be achieved. Mathematical modelling of the digestion process for the first time has given a better understanding of the gas production process, which has increased gas yield from the same amount of sludge. Monitoring and targeting of the power output, together with tighter maintenance regimes, have helped to reduce plant down time.

In addition to getting more out of the existing plant, the NFFO and the ability to sell power to the Grid (rather than simply using it on site) has enabled Thames Water to maximise the financial benefits that can be derived. At works in Reading and Slough, for example, dual fuel compression ignition engines were dismantled 15 years ago because they were proving uneconomical. With the advent of

NFFO, both these sites have been fitted with new spark ignited engines which produce around 400 kW of electricity at each site, from gas which would otherwise have been flared off.

THE FUTURE

A new sludge disposal alternative being considered by many water companies is incineration by fluidised bed techniques.

In Thames Water we plan to have two such incinerators operational at the two largest sites - Beckton and Crossness - by 1998.

Whereas most sewage incineration plants being built currently have no power recovery equipment, at Thames Water we made the decision to invest in the additional cost of the power plant and therefore each incinerator will be equipped with a 6 MW steam turbine power recovery facility. This adds considerably to the complexity of the plant, which is the reason most operators shy away from this option. However, Thames Water believes its decision demonstrates its commitment to the environment and in the long term will benefit the company economically.

It is unfortunate that due to the exclusion of sewage schemes from the current NFFO rounds, electrical generating capacity provided in this way will not be used by the government in meeting its renewable energy targets for the year 2000.

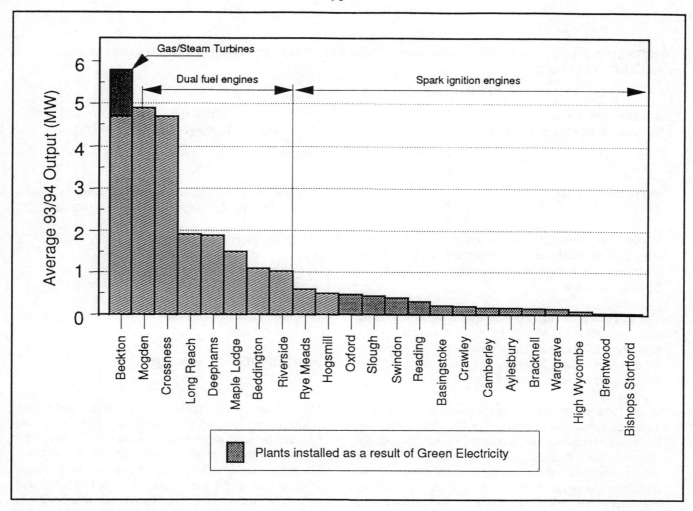

Figure 1: Range of Plant Sizes

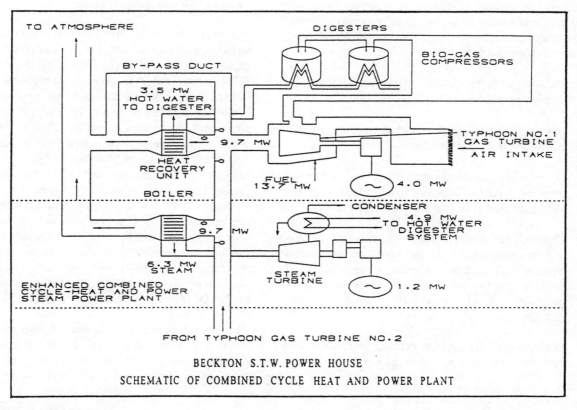

Figure 2: Schematic of CHP plant

UPGRADING LONDON UNDERGROUND'S GENERATION TO ACCOMMODATE THE CHANGING PATTERN OF LOAD

A S Jhutty, S R Carter
Kennedy & Donkin Power (UK)

D C Knights
London Underground Limited (UK)

INTRODUCTION

A significant milestone was reached last year in the history of electricity generation in the UK with London Underground's Lots Road Power Station reaching its Ninetieth birthday.

The power station has been instrumental in providing safe and reliable power to LUL's electric trains throughout this period. The power station is a testament to the skills and ingenuity of the electrical and mechanical engineers involved.

In 1994 the power station operators noticed a trend of increased variations in demand and system frequency. Excessive variations in frequency have undesirable effects on plant operation and, in the case of LUL, there are also serious safety implications as some of the signalling equipment is supplied by rotating frequency converters. The trend was suspected to correlate with the phased introduction of new rolling stock.

Initial measurements indicated poor load sharing between the generators (Figure 1). Turbine Alternator No. 5 (TA5) is seen to be the most responsive. It was therefore decided to investigate the problem by adopting the following procedure:

(a) Measurements to quantify plant performance by monitoring electrical and steam quantities

(b) Develop computerised dynamic models of the generating units and the electrical plant

(c) Develop computerised probabilistic model of each type of traction equipment and determine the aggregate load as seen at the power station

(d) Undertake studies to evaluate the effect of the new traction equipment and the effect of anticipated changes. The latter includes greater usage of modern rolling stock with regenerative braking capability which accentuates sudden power changes.

(e) Determine limitations of existing generation and possible corrective measures to improve response

time, load sharing and overall system frequency stability both for the present and the future load

(f) Apply corrective measures

(g) Verify the analysis with measurements following the completion of the corrective measures

This paper outlines each of the above aspects.

DESCRIPTION OF ELECTRICAL SYSTEMS

The London Underground has been generating electricity for traction purposes since 1890 when the City and South London Railway opened from Stockwell to King William Street. It was decided to centralise generation at a new power station at Lots Road. Construction of the power station started in 1902 and was described by the "Illustrated London Mail" in September 1904 as the largest powerhouse ever built. Electrically powered trains started operating in 1905. The present power station equipment, resulted from the 1965-69 fourth major replanting.

The plant consists of six two cylinder Parsons steam turbines operating at 3000 r.p.m. coupled to alternators capable of a maximum continuous 30MW at 22kV and 50Hz. Steam for the turbines is supplied from six Babcox & Wilcox vertical membrane wall boilers rated 136000kg of steam per hour. They supply steam to the alternators via a common steam range at 482°C. The output of a single boiler is approximately equal to that required by a single turbine. Although originally designed to burn heavy fuel oil, economics of fuel pricing lead to conversion to dual-fuel firing capability in the late 1970's. Natural Gas is now the main fuel, but the boilers can be changed over whilst "on-line" to operate on light distillate oil. This facility gives great flexibility and enables advantage to be taken of cheaper interruptable gas tariffs. Oil is also used for short periods whilst undertaking certain operations on the boilers for convenience and safety.

The power generated at Lots Road is distributed at 22kV via a ring of cables to switchhouses at five strategic locations around London at which, the power is transformed down to 11kV and fed to traction

Opportunities and Advances in International Power Generation, 18–20th March 1996,
Conference Publication No. 419, © IEE, 1996

substations. The substations are situated alongside the railway at intervals of about 4 km. However, a few substations are fed at 22kV to minimise distribution losses. At each substation, the power is converted to 630V d.c. to supply traction power to the trains. Substations also provide auxiliary supplies for lighting and signalling purposes. A full description of the LUL Power System has been given by Bletcher (1).

A second power station is located at Greenwich which is utilised for peak lopping and as a standby to Lots Road. This station contains seven aero-derivative simple cycle gas turbine alternators each with an ISO rating of 14.7MW, although the economic operating point of these generators is 11MW. These units are designed to be fully automatic and are controllable from Lots Road Control Room from where they can be run up to give full loading in about 2 minutes (emergency rating). As at Lots Road the machines can be operated on either Natural Gas or Light Distillate Oil.

The two power stations provide about two thirds of the total power demand, the remainder being supplied by three Bulk Supply Points from the regional electricity company. Two further Bulk Supply Points have recently been constructed which will enable all of London Underground's electricity supplies to be bought in and the power stations used only for emergency supplies. As further enabling works are required before the Bulk Supply Points can be utilised, it is likely that LUL will continue to generate electricity until at least 1999.

EVOLUTION OF TRACTION LOADING

Traction equipment is one of the most disturbing loads to be supplied by a power system. The load is by its nature highly variable, not only does the requirement for power change throughout the day as the number of trains in service changes, but the loading can also change very suddenly.

Prior to the introduction of the new Central Line trains, all London Underground rolling stock have been controlled using Pneumatic Camshaft Motors (PCM). These control the acceleration of the train d.c. motors by switching out resistors and changing the configuration of the pair of motors in the circuit until full parallel running is achieved. The 'notching' is under the control of a sensing relay, which gives a nearly constant current characteristic until full line voltage has been applied, at which point the back-emf of the motor controls the current flow. This type of train will operate at reduced power output if the line voltage is reduced.

The new Central Line trains utilise thyristor controlled drives to give a constant power output and hence more predictable and higher acceleration than the previous units regardless of traction line voltage. This results in a faster train service and reduces the overall number of trains required for the service. The new trains are also provided with facilities for full regeneration of braking energy into the track supply system.

The new trains have peak power demands of 2MW compared with approximately 1.5MW for the traditional units. This feature, coupled with the ability to regenerate up to 2 MW into the power supply, results in a highly peaking demand profile.

During the next three years the further changes will take place in the LUL rolling stock, all of which will have an effect on the power station. The Central Line trains will begin to operate with full performance and the Northern Line and the Jubilee Line will also be eqipped with new trains similar to those operating on the Central Line.

It was considered important to evaluate whether or not the changes in loading type could cause instability in the power system. There is an effective lower limit on frequency of 47.5Hz because of protective devices on the Victoria Line signalling systems. If the frequency drops to 45Hz, the power station itself could become unstable and loss of supplies to the railway would occur. Sudden decreases in demand are less de stabilising, but may cause problems with an excess of steam in the boilers.

DESCRIPTION OF LOTS ROAD GENERATOR GOVERNING SYSTEM

The prime movers are driven by high pressure steam produced in the dual-fired boilers. The steam is delivered to a steam chest from where it enters the turbine via two governor valves P1 and P2. Valve P1 is the main governor valve which admits steam to the first stage (governor stage) blading. P2 valve admits high pressure steam to the turbine downstream of the governor stage blading to provide overload capacity. Since this reduces the turbine efficiency, P2 is only activated during high demand conditions (i.e. above a nominal 73% of unit rating). The tendency is to operate the turbo alternators with P1 fully open and therefore close to the point at which P2 begins to open and, due to differences in the governor settings, the load is not equally shared between the units.

During load acceptance tests, the steam pressure in the inlet valves was observed to vary only slightly, indicating a stiff system with adequate capacity, pressure and steam velocity. Thus, the time constants of the the governing system are dominated by the governor and the associated valve operation.

TESTS AND DYNAMIC ANALYSIS OF THE SYSTEM

In order to assess the magnitude of power swings which could be supported by Lots Road before the frequency fell below 47.5Hz the generation and steam system at Lots Road was modelled using UMIST's Interactive Power System Analysis (IPSA) software. To validate the governor and turbine models, step load-application tests were carried out on TA5 and TA6 at Lots Road.

An 11MW load acceptance test was performed at 00:30 hr before the steady night load was reached (Figure 2). TA5 and TA6 were generating 21MW each, GT1 at Greenwich was generating 11MW and GT2 was generating 3MW. GT2 was held at maximum output on load limiter control so as not to interfere with the tests, while providing some spare spinning reserve in the event of the tripping of one of the Lots Road sets. GT1 was tripped and the load picked up by TA5 and TA6.

Measurements were taken of the voltage, frequency and total power at Lots Road, the standby gas turbine power, the P2 steam pressure, the governor pilot oil pressure, the P1 steam pressure and the emergency valve steam pressure. After a few seconds of swinging the load was shared by TA5 and TA6 according to their different governor and steam valve characteristics. P2 valve was clearly seen to operate during the test for TA5 but did not appear to operate for TA6.

A second test was undertaken an hour later under a little more load than normal night time conditions in order to observe the response of the sets away from the range of operation of their P2 valves. Although both P2 pressures were seen to change, this was probably caused by the back pressure from the turbines.

A governor and turbine model was developed and included simulation of the operation of the P1 and P2 valves. The models for TA5 and TA6 were validated against the test results.

The output of the simulation was compared with previously recorded events. One of the recordings was for a 35MW change on a pre-disturbance load of 83 MW, which occurred on 9 November 1994 at 15:35 over a period of 54 seconds with 5 units in operation (Figure 1). The simulation (Figure 4) compared favourably with the actual machine responses. The output of the simulation of a 25MW change on a pre-disturbance load of 77 MW which occurred on the same day at 20:30 over a period of 17 seconds with 4 units in operation also compared favourably with the actual machine responses. Thus the simulation was taken to be validated by these studies.

The model was used to determine the effect of progressively increasing the pre-disturbance load and the swing magnitude until the frequency fell to 47.5 Hz. The analysis was repeated with idealised governor and turbine models to determine the benefits associated with a balanced system. The results indicated that the differences in performance of the individual machines are significant as the severity of the load swings increases. Once the sets hit their output limits they cease to provide any further help in controlling the frequency. Thus, although the frequency may be subsequently restored the depression will be deeper than in the balanced case.

Figure 5 indicates that with 5 units operating at a pre-disturbance load of 83 MW, a 78 MW load change would cause the frequency to fall to 47.5 Hz with the existing governors. However, the frequency fall would be limited to 48.5 Hz if the governors were balanced.

PROBABILISTIC MODELS OF TRACTION LOAD

The number of trains starting, motoring, stopping, coasting and at standstill at any time can be considered to be a random process; thus the power demand of the trains can be considered to be probabilistic. A computer model was developed to model the power demand of the rolling stock and to assess the effect of regenerative braking on the total load demand at Lots Road.

The total number of trains required for peak service is 460. It is estimated that during peak times about 336 of these take their power from Lots Road and Greenwich Power Stations.

A typical train will accelerate on leaving a station and reach a coasting speed. The simplest journey is finished by braking into the next station. Braking and accelerating can occur intermittently throughout the journey depending on a number of factors such as the signals, speed limits, curvatures and gradients. Based on recordings of a Northern Line 1972 (NL 72) stock camshaft train in September 1994 several train loading patterns were modelled for a period of 700 seconds.

The trains on the Central Line were modelled with their higher power requirements and regenerative characteristics. The future regenerative Northern Line trains were also modelled.

A random start time between 0 and 400 seconds was assigned to each train and the total demand at Lots Road obtained, taking account of the generation at Greenwich and a base load of 10MW for the non traction load supplies. The model was validated by comparing snapshots of the total power measured at Lots Road.

Several simulations were carried out with a mix of camshaft trains and the new trains. The probablistic analysis utilised a seeded random number generator to perform 1000 iterations of each scenario to obtain the probability density function of the maximum positive change obtained.

The cumulative distribution function (cdf) of the probability of different magnitudes of load change occurring within a five minute peak period was calibrated against known load changes (Figure 3). It was found that the 97 percentile value corresponded with actual recorded load changes. The simulation confirms that the increased load changes experienced at Lots Road power station are of a magnitude consistent with the operation of the regenerative Central Line trains. Based on the 97 percentile value the model predicts that the surges will increase to the order of 75MW in the future with the new Northern Line trains.

METHODS FOR IMPROVING SYSTEM PERFORMANCE

At the time of writing, the new Central Line trains were temporarily operated on reduced performance and reduced regeneration because of system constraints. One of the constraints was the risk of excessive frequency deviations. Clearly this is unsatisfactory in terms of the business objectives and also for energy conservation, costs and loss of revenue.

The salient options for improvements are:

(a) Adjust governors to ensure better load sharing

(b) Replace the mechanical governors with fast and precise electronic governors

(c) Replace governing valves with faster single stage valves and install electronic governors

(d) Run gas turbines at Greenwich during the day to reduce the total load and increase the spinning reserve.

(e) Introduce a new source of power or enhance existing facilities (eg the grid)

The studies above indicated that option (a) would be the most cost effective solution in the short term. Based on the above analysis, the maximum load change expected with the Central Line trains operating at full performance is 60MW and this can be accommodated by the existing system.

It should be noted that long term investment is not considered desirable as the power station is expected to close in 1999. However, with the introduction of the new Northern Line trains the total power consumption will increase and the power changes are expected to increase to 75 MW. The other options would be more costly and need to be considered in the overall context of future load growth and future power supply arrangements including emergency power requirements.

CONCLUSIONS

This paper serves as a reminder of the need to provide generation plant and controllers suitable for the requirements of the load. Ideally the complete system including generation, distribution and load should be optimised in terms of performance, energy efficiency, cost effectiveness and safety. The paper has also demonstrated the usefulness of probablistic analysis when applied to complex phenomena on such an integrated system.

In this case the introduction of more powerful new technology regenerative trains adversely affected the operation of the generators. The problem will be overcome in the short term by adjusting the settings of the existing mechanical governors to improve the load sharing between the machines. This will increase the ability of the power station to respond to sudden changes in demand which have worsened with the change in characteristics of the Central Line trains. The introduction of new rolling stock on other lines will require additional improvements.

REFERENCES & ACKNOWLEDGEMENT

1. Bletcher, M. I., 1987, "Electricity supply system on the London Underground", Power Engineering Journal, 1987, 317-324.

The Authors would like to thank the Engineering Director of London Underground Limited for Permission to publish this paper.

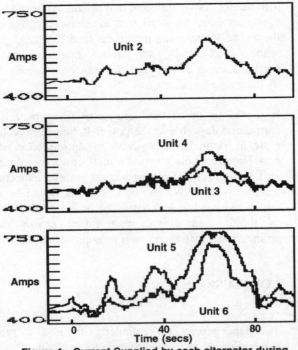

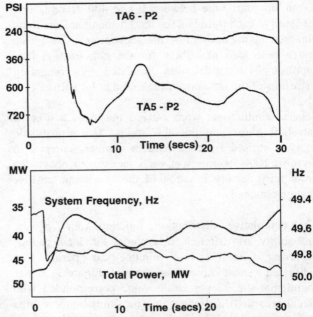

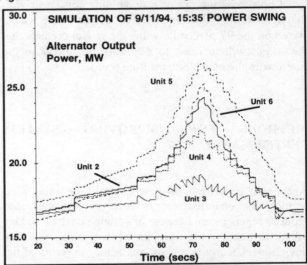

Figure 2   Measured Quantities during load application tests

Figure 1   Current Supplied by each alternator during an actual load change

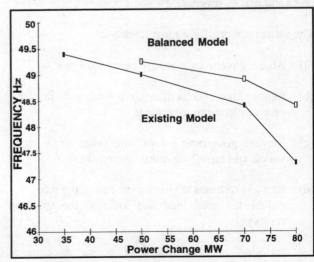

Figure 4   IPSA Simulation of Figure 1 Load Change

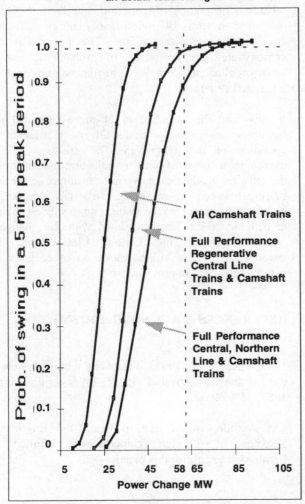

Figure 3   Cumulative Distribution Function of Power Change

Figure 5   Frequency Response with existing (unbalanced) and balanced governing. Pre-disturbance Load 83MW, 5 Units.

NUCLEAR ENERGY'S ROLE IN A MORE SUSTAINABLE FUTURE

WILLIAM TURNER AND JEREMY WESTERN
NUCLEAR ELECTRIC PLC, UNITED KINGDOM

Introduction

Increased population growth and economic development are accelerating the rate at which energy, and in particular electrical energy is being demanded. All methods of electricity generation have consequences for the environment so meeting this growth in demand while safeguarding the environment poses a growing challenge.

Many nations, now conscious of the impact of industry on the environment, are seeking ways to preserve more of the world's natural capital for future generations. There will always be a conflict between the provision of vital goods and services, such as electricity, and the implications for the environment. Energy is an essential element in the debate about sustainable development and how to achieve it. There is a strong correlation between people's standard of living -measured by gross national product per capita - and energy consumption. Informed decision-making can lead to a more sustainable energy future and a cleaner environment. This paper examines how the world's voracious demands for energy may be made more consistent with this concept of sustainable development. In particular the role that nuclear power might play in filling the growing gap between what the world wants to consume in terms of energy and what the environment tells us we can sustain is considered.

World Growth in Electricity Demand

As society developed [from primitive man to technological man] so energy consumption per head has risen. The 'high tech' man of today consumes about one hundred times as much energy as primitive man. Over the last century there has been a worldwide shift from agricultural based economies to industrial and 'high tech' economies. The resulting increased demand for energy has been met by the fossil fuels - initially coal, but lately more by oil and gas. Coal accounts for 27%, oil for 40% and gas 23% of current world energy supply. Latterly nuclear power has made a contribution [7%], hydro has a mere 3%.

Growth in energy demand is the product of two separate factors. Firstly, the increase of use per capita driven by the societal changes and secondly.

by the growth of the global population itself. The UN predicts that world population will increase from the present 5.3 billion to over 8 billion by 2025.[1] That increase will take place almost entirely in the continents currently undergoing industrial transformation - i.e. Africa, China, and India. This means that world population is increasing at a rate equivalent to adding the population of the UK every six months! Primary energy use varies greatly around the world. The industrial countries which have just 20% of the World's population consume around 70% of its energy. The North Americans consume per capita more than 5 times the world average and more than 10 times the average in the developing countries.

At least in the OECD countries energy consumption since the 1970s oil crisis has effectively levelled out as Figure 1 shows. Energy growth has been slower than GDP reflecting a decrease in the 'energy intensity' of the economy. Figure 1 also shows that although energy growth has slowed, growth in electricity demand has continued and is now twice the level in 1970. In other words electricity is steadily taking a bigger share of the energy market. A different pattern exists in the developing countries 9 (Figure 2). Here energy growth matches that in GDP and has doubled in the last twenty years. Even more striking is their growth in electricity consumption, bearing out the strong effects of the increased urbanisation of the developing countries.

According to the World Energy Council, we can expect a doubling, or even tripling, of global electricity demand by 2020. Electricity demand in the OECD countries will also continue to grow but at a much slower rate, most of the growth will be in the developing countries whose consumption is projected to rise two to three times that of the developed world.

Figure 3 shows the fuels used to generate electricity in the OECD countries in 1973 and in 1990. The big change has been the replacement of oil burning capacity by nuclear power which currently generates around a quarter of all OECD electricity. In future electricity from coal is expected to decline at the expense of a rise in gas fired electricity; while nuclear is projected to stabilise at about its present market share.

Opportunities and Advances in International Power Generation, 18–20th March 1996,
Conference Publication No. 419, © IEE, 1996

But is this situation sustainable. Sustainable development does not mean having less economic growth. Quite the opposite, a productive economy is able to generate the resources for environmental protection and improvement. Investment is essential if environmental improvement is to be realised. It is important to recognise that not every aspect of the present environment needs to be preserved for sustainable development to be achieved. Instead, it requires due consideration of environmental impacts to be taken into account throughout society in all its decisions. Decisions by the process operator, government or public must be based on the best possible scientific information and analysis of the risks. Where there is uncertainty and the consequences of a decision are potentially serious, precautionary action may be necessary. This raises a number of questions in the way we currently produce and use energy.

The Greenhouse Effect

One problem in particular where these issues are having to be addressed is that the combustion of coal, oil and natural gas results in the production of significant quantities of carbon dioxide (CO_2), which is the principal contributor (72%) to the greenhouse effect. Greenhouse gases, such as CO_2 and methane, reflect heat back to earth and in the process warm the lower atmosphere and the Earth's surface. Globally human activities are causing the concentration of these gases to build up in the atmosphere at an ever increasing rate. For example CO_2 concentrations in the atmosphere are rising at a rate of around 0.5% per annum.

Emissions of CO_2 over the last century follow the growth in fossil fuel energy consumption. This continuing increase in greenhouse gas emissions presents a serious threat through global climate change unless the trend can be reversed. The resulting rate of temperature rise is forecast to be about 0.3°C per decade. The Intergovernmental Panel on Climate Change's (IPCC's) latest assessments[2] are that temperatures will increase by about 1°C above present levels by 2030 and 1.8°C by 2060.

In the widely leaked draft of the latest IPCC report[3] there appears to finally be a consensus amongst the scientists involved that there is a detectable human influence on the global climate and that global warming could have dramatic effects. 1995 was one of the hottest years on record. Nine out the ten warmest years have occurred since the early 1980s. The warming has resumed during the last two years after a brief lull caused by the volcanic eruption of Mount Pinatubo in 1991. The IPCC now warn the

consequences of climate change could be that tropical diseases could increasingly spread into temperate areas, that droughts and floods may increase, that forests may die out and that harvests in poor countries could fall dramatically. Insurance companies are already looking at forecasts that certain extreme meteorological events will occur more often, conscious that the insurance market is already carrying heavy losses following a series of natural disasters including storms across Northern Europe and the hurricanes Hugo and Andrew in the USA.

In recognition of the threat of global warming and as a signatory to the UN Climate Change Convention, the UK government has committed itself to returning its CO_2 emissions to 1990 levels by the year 2000. This can only be seen as a first step given that the IPCC has calculated that emissions must be reduced by at least 60% to achieve stabilisation of atmospheric CO_2 concentrations (Figure 4). Switching from coal to gas generation of electricity, as is occurring in the UK, brings the twin benefits of increased thermal efficiency from Combined Cycle Gas Turbines (CCGTs) and the use of a fuel with a lower carbon content per unit of heat content. In the UK this change is occurring rapidly, chiefly as a result of commercial interests. Table 1 shows that it is this switch to gas for electricity generation combined with increased output from the country's nuclear power stations which has put the UK in the position of being one of the few countries likely to meet its 2000 CO_2 emission target. This will be achieved despite the current failure of emission savings schemes put forward in the national Climate Change Programme. However the potential reduction in emissions from the change to gas as a fuel is limited ultimately by the CO_2 emission rates from CCGT plant. Gas fired generation still releases half the CO_2 emissions of coal.

Other studies, such as that by the International Project for Sustainable Energy Paths (IPSEP)[4], suggest that world CO_2 emissions should be reduced to 20% of present levels, over the next 100 years, to ensure average temperatures do not increase by more than 2°C above present levels. This might be done by setting a cumulative emission limit of 300 GtC, possibly split 50/50 between the industrialised and developing nations (Figure 4). However, this difficult adjustment cannot be imposed equally on the developed and underdeveloped world, since that would imply stopping growth in its tracks in the poorest nations. Instead some sharing of the burden of adjustment that allows growth to continue in both the developed and the less developed world needs to be agreed. This would require industrial countries to make an early and fundamental adjustment. Even so, a period of, perhaps, 100 years is probably necessary to complete the process of such a major change.

Increasingly, it is being seen as the role of the industrialised countries to help the developing nations take a 'short-cut' to efficient use of energy. This has links with the concept of 'joint implementation' discussed at the 1995 Berlin Conference of the signatories to the climate change convention. Joint implementation envisages industrialised countries investing in efficient energy use in the developing world in return for carbon dioxide emissions credits "at home".

If the World is to stand a chance of meeting the more demanding CO_2 emissions targets - such as those suggested by IPSEP and echoed in Berlin this year - it will require the combined strategy of a switch to lower CO_2 emitting fossil technologies (such as CCGTs) together with increased use of "CO_2-free" technologies (such as nuclear power and renewables) together with implementing energy conservation measures to the maximum. Over the last 30 years nuclear power has been one of the major factors that has helped limit the per capita level of CO_2 emissions. Electricity generation from the nuclear fuel cycle results in only 4% of the CO_2 which arises in the generation of electricity from coal. Globally, nuclear power sources save 490 million tonnes of carbon emissions per year.[5] That is approximately 1 tonne of carbon for every 10 people on the planet every year. The UK's nuclear power stations currently save the emission of around 18 million tonnes of carbon each year compared to if the same amount of electricity were generated by the existing UK fossil fuel plant mix. Official projections indicate UK CO_2 emissions starting to rise again beyond the year 2000. This rise occurs partly because they assume no further nuclear construction post-Sizewell B. This would lead to a gradual but inexorable reduction in nuclear generation as older power stations are retired from service. If nuclear power were to be phased out, these savings would become realised emissions.

Acid Rain and other Pollutants

There are also major concerns about the problem of acid deposition from the combustion of fossil fuels. This occurs mainly through rainfall, containing sulphuric and nitric acids in solution, in turn caused by emissions of sulphur dioxide (SO_2) and nitrogen oxides (NO_x) in exhaust gases. While NO_x is produced in all combustion processes, SO_2 is produced only in the combustion of fuels containing sulphur, principally coal and oil. The electricity supply industry is responsible for about 70% of the UK's SO_2 emissions, and 30% of its NO_x emissions. The impacts of acid rain include effects on human health, buildings, forestry, fisheries and natural ecosystems.

The ecological impacts of acid rain depend on the location in which it falls. The effects can be long lasting and are not necessarily in proportion to the amount deposited. This has led to the development of the concept of 'critical loads' to measure the sensitivity of different areas to the harmful effects from these pollutant gases. The Department of the Environment[6] defines the critical load as "the quantity of a substance falling over a given period below which a specified part of the local environment can tolerate without adverse effects occurring". Many areas in the UK have already exceeded their 'critical load' for acid deposition and the damage this has caused is evident.

The United Nations Economic Commission for Europe (UNECE) has adopted the critical loads approach as part of its revision of the protocol for control of emissions of sulphur which is due to be signed in June. The new reduction targets for the UK of 70% in 2005 and 80% in 2010 (compared to 1980 levels) are very stringent because of the wide area of particularly sensitive ecosystems UK emissions affect. The delay in implementation of the targets to 2010 amounts to a delay in application of the "polluter pays" principle. Even after 2010, these reductions will still allow some of the more vulnerable areas to be subject to deposition greater than their critical loads. This implies that a degree of damage is acceptable.

Sulphur dioxide emissions can be reduced from coal fired stations by fitting flue gas desulphurisation (FGD) systems to remove the gas from the exhaust. This process is costly and not without its own consequential environmental penalties such as the need for quarrying large quantities of limestone, which in the UK tends to be found in landscapes of high scenic value. Another problem is the disposal of very large amounts of the gypsum by-product and the liquors containing heavy metals. The fact that retro-fitting of FGD reduces efficiency also leads to increases in the emissions of CO_2 per unit of electricity generated.

As a non SO_2 emitting form of generation nuclear power is playing a major role in helping the UK meet its international commitments to cut sulphur emissions and so reduce the problem of acid rain.

Control of NO_x emissions is also complex. New burner technologies have been developed which reduce the quantity of NO_x produced in fossil fuel combustion, but significant quantities are still formed. Technologies to remove NO_x completely are far more expensive. Although gas fired generation emits less NO_x than coal and oil fired plant, NO_x emission levels from CCGTs may still be a problem in the future where they contribute to ambient NO_x levels that approach health air quality standards. By not emitting NO_x nuclear generation is contributing to improving air quality

Renewables

Renewable energy sources add to the diversity of supply and, like nuclear, avoid significant CO_2 emissions. The contribution the renewables can make over the next few decades should be maximised but it should not be unrealistically overstated. The World Energy Council[7] suggests that "even given clear and widespread public policy support, the 'new' renewables will take many decades to develop and diffuse to the point where they significantly substitute for fossil fuels". The risks of getting the mix of generating plant wrong appear higher now than at any previous time. Maintaining a significant proportion of nuclear power plan that can be built to schedule provides a valuable safety net.

Renewable energy technologies advocated for the UK include wind power, waste burning and the burning of landfill or sewage farm off-gases. Although there are many potential sites for renewable energy schemes, these are limited by a range of practical problems, many of which are environmental in nature.

The main external environmental impact of wind power is visual intrusion and local noise, together with the indirect environmental costs due to the energy and resource input within each unit's construction. In the UK the best wind generating sites are typically found in exposed areas on hills and coasts; these are frequently areas of natural beauty such as the Yorkshire Dales. Sensitive siting can reduce this visual impact but, due to the diffuse nature of the resource, large areas must be dedicated to wind farms if a significant contribution is to be made by this technology. For a wind farm to produce 1260MW, the same output as the Sizewell B PWR reactor in Suffolk, an area of greater than $500km^2$ would need to be covered with wind turbines.

Combustion of renewable fuels, can provide electricity without the net emission of greenhouse gases. However, in common with the combustion of fossil fuels the combustion of renewables materials (biomass) does release oxides of nitrogen, and in the case of waste burning, sulphur dioxide, hydrogen chloride and heavy metals are also emitted. Hydrogen chloride is an acid rain gas acting in a similar manner to NO_x and SO_2, whilst heavy metals are often highly toxic to plant and animal life. The effects of heavy metals can be made worse by the process of bio-amplification which allows these chemicals to be concentrated at ever higher levels as they pass through the food chain. Control of these so-called 'air toxics' is currently receiving particular attention in the US.

Nuclear Power

The World Energy Council[7] recently concluded that, subject to the resolution of several issues of public concern, a considerable expansion of nuclear generation may be assumed by 2020 to help meet growing worldwide energy demand on a sustainable basis. The UK has a strong technology base and experience in nuclear power to allow it to benefit from this opportunity for growth in a sustainable source of non-polluting energy.

There are a number of significant environmental benefits arising from the use of nuclear power, but it does raise its own environmental issues. During operation some radioactivity is released at a very low level into the environment either via filtered emissions to the atmosphere or in liquid form in the cooling water discharged to sea. In the UK these discharges are strictly controlled by the Government regu.atory authorities.

The levels authorised to be discharged by the UK nuclear industry are such that the additional radiation dose to the public is extremely small in comparison to that due to natural background radiation. By measuring and controlling discharges to the environment through the use of filtration and ion-exchange treatments as necessary, the power station operator ensures that all releases are below authorised levels. Each facility carries out and publishes results from its own monitoring programme on the effects of its routine discharges and this is backed up by independent measurements made by the regulatory bodies. These measurements, carried out since before the stations were even built, and have shown that in general there has not been any build up of radioactivity in the surrounding environment due to their operation.

Although undoubtedly an issue of public concern and desptite the requirement for several decisios still to be made relating to its ultimate disposal, the significance of radioactive waste from the nuclear fuel cycle must be considered in perspective against the much greater environmental impacts from the wastes of other fuel cycles. The volume of the radioactive waste produced in the UK amounts to less than 1/100th of the toxic wastes produced by UK chemical industries[8]. Moreover, contrary to public perception, many of these toxic chemical wastes remain hazardous for longer than radioactive waste. All the costs associated with the safe management of waste disposal are provided for by the nuclear generators and so are included in the cost of nuclear electricity.

At the end of its economic lifetime each nuclear power station must be decommissioned so that ultimately the site can be returned to alternative uses without any need to account for its history as a nuclear site. Until this stage has been reached, the operating company remains responsible for the station's decommissioning which must be conducted in such a way as to ensure the continued safety of the public, the work force and the environment.

The process of decommissioning is by no means unique to nuclear power. Some chemical companies, for instance, face considerable challenges. For these companies many serious contamination problems remain unresolved, not least of which is the financing of clean-up operations. Few industries are as thorough as the nuclear industry in planning for decommissioning, and making financial provisions to pay for it.

The process of decommissioning is well understood. Around 80 reactors have come to the end of their economic lives, and several sites have been completely cleared. The same high standards apply to the handling and disposal of decommissioning waste as to routine waste from power generation.

Environmental Economics and Sustainable Development

As fossil fuel resources are depleted there will be a move to more marginal resources which require more energy for their extraction and so will be associated with greater environmental impact. As they become more scarce substances such as oil and gas may be seen to be more valuable as chemical feedstocks than as fuels to be burned for electricity generation. There is a growing call for countries to consider their response to this challenge of increased "sustainability" today. The use of nuclear power is one of the factors that can play a significant role in aiding sustainable development. Like most resources the uranium used to fuel nuclear reactors exists in finite quantities. Known low cost reserves of uranium are almost 4 million tonnes. This is sufficient to meet current needs for around 60 years. But the International Atomic Energy Agency estimate up to 4 times this amount is yet to be discovered. Unlike fossil fuels uranium is a high energy content resource which has no other major use and is unique in that uranium and plutonium can be recycled from spent reactor fuel and used as Mixed Oxide fuel, extracting 50% more energy from the original uranium resource. There is also the potential technology of the fast reactor. This has the capability of realising up to 60 times more energy from the original uranium than is possible from thermal reactors. With the use of fast reactors, uranium reserves would be nearly twice the value of the world's total reserves of all fosil fuels.

The UK Government[9] believes that its objectives for both energy and environmental policies can best be achieved through the mechanisms of the market. It has stated that the market is the most effective and efficient means for meeting energy needs, both nationally and internationally, but it recognises that the market cannot give proper weight to environmental considerations unless the costs of environmental damage or the benefits of environmental improvements are built into the prices charged for goods and services.

There is growing support for moves towards 'full cost pricing'. In this way market forces can be harnessed to deliver environmental and social benefits in a way which maximises economic efficiency. Explicit monetisation of a wide range of environmental impacts allows the specific damage per unit of energy produced to be expressed. This permits direct and consistant comparisons between conventional internal costs and these external environmental costs so that the benefits and costs of the various fuel cycles can be assessed on equal terms.

The existence of difficulties is acknowledged, such as how to proceed if the external damage cost is not known or how to deal with the damage costs of residual discharges, if conditions are close to optimum. Despite these problems, various authors indicate that, quantification and monetisation of external environmental costs is likely to lead to better quality decisions than a zero value and that, on balance, society is likely to be worse off if external costs are not used in the decision process than if they are used[10].

Conclusion

"Sustainable development" requires continued growth in energy supply in the developing world, together with a determined effort by the present industrialised world to introduce energy efficient processes and to put an appropriate value on fuels to take proper account of the full costs of their use. Environmental constraints and to a lesser extent, the dwindling of resources, is expected to force a reduction in the use of fossil fuels and a corresponding increase in renewables and nuclear power. The developing nations like India and China must be expected to utilise their indigenous fossil reserves and this will require the present industrialised world to compensate by burning less fossil fuels such as gas and switching to non-fossil fuel methods of generation. As an existing readily available source of electricity free from CO_2 emissions, nuclear power has a key role to play in providing a vital bridge to a sustainable energy path. If this path can be found sustained economic growth can perhaps be achieved whilst also meeting acceptable environmental constraints.

References

1   United Nations. World Population Prospects. 1992.

2   Intergovernmental Panel on Climate Change. Climate Change 1992: the IPCC supplementary report. Cambridge. 1992.

3   The Independent on Sunday, 15 October 1995

4   Krause F, Bach W. & Koomey J. Energy Policy in the Greenhouse, IPSEP, Earthscan Publications, London 1991.

5   The Uranium Institute. Nuclear Power, Energy & the Environment. 1992.

6   NERC. Critical Loads: concept & application. HMSO. 1993.

7   World Energy Council. Energy for Tomorrow's World. Kogan Page. 1993.

8   British Nuclear Fuels plc. Nuclear Waste: what's to be done about it? 1989.

9   Department of the Environment. Sustainable Development: the UK strategy. HMSO. 1994.

10   Woolf T. Its Time to Account for the Environmental Costs of Energy Sources, Journal of Environmental Management, 1993

Table 1

Monitoring of UK CO_2 Emissions (million tonnes carbon per year)

Years	1990	1991	1992	1993	1994
Change relative to 1990 due to:					
Changed energy efficiency		+ 1	+ 2	+ 1	- 3
Changed GDP		- 3	- 4	- 1	+ 5
Weather		+ 5	+ 3	+ 2	+ 1
Sub-total		+ 3	+ 1	+ 2	+ 3
Changes in electricity generation		- 2	- 4	- 9	- 13
Total annual emissions	158	159	155	151	149

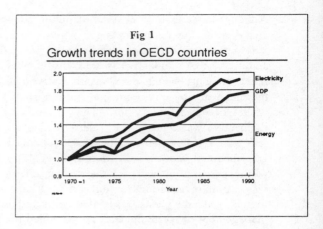

Fig 1

Growth trends in OECD countries

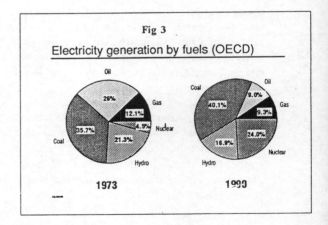

Fig 2

Growth trends in developing countries

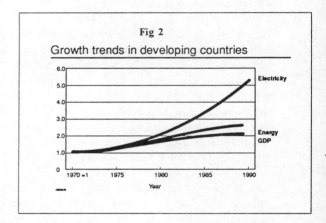

Fig 3

Electricity generation by fuels (OECD)

1973   1990

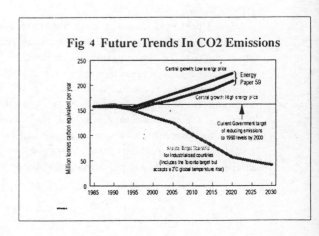

Fig 4 Future Trends In CO2 Emissions

THE 'SUCCESS' OF THE COMBINED CYCLE GAS TURBINE

W J Watson

Science Policy Research Unit, University of Sussex, UK

Over the last 20 years, a new technology for power generation has emerged. The Combined Cycle Gas Turbine (CCGT) has been installed in increasing numbers throughout the world. Synthesised from traditional gas and steam turbine technology, the CCGT appears to have been at a sufficiently mature stage of development to take advantage of the worldwide shift towards a more 'market-driven' economic climate. Also of importance has been the availability of large quantities of natural gas, coupled with widespread concern at the environmental effects of traditional technologies.

For reasons that owe a lot to the political agenda of the Thatcher Governments, the trend towards the CCGT has been most evident in the UK. The 'Dash for Gas' that followed the privatisation of the Electricity Supply Industry has had wide ranging effects on the whole energy industry. It was the particularly devastating effect of privatisation and the 'Dash for Gas' on the UK's coal mining industry that led to this investigation of this new and apparently invincible technology. Its sudden appearance was met with surprise by many people who had never heard of a CCGT. There were questions to be answered. Where did it come from and who developed it ? This was the starting point of the research. In an age when nuclear power is no longer an option in most countries, how has the CCGT come to be in a position to out-perform all other competitors, including many promising 'clean coal' technologies such as Fluidised Beds ?

In this short paper, the aim is to re-evaluate the conventional wisdom that the CCGT has become so popular solely due to changed market circumstances and cheap natural gas. It will be demonstrated that many political and technological factors have had a part to play, particularly in countries such as the UK, where the CCGT has had a large impact. To do this, the history of the CCGT will be analysed from three related perspectives. These will examine the technology itself and its attributes, the role of the equipment manufacturers, and the involvement of Governments. The conclusions will hopefully lead to some wider debate about our general view of technological development, how it should be handled and by whom.

THE HISTORY OF THE CCGT

As part of the analysis of its apparent success, it is necessary to give a short history of CCGT technology, with particular reference to its prime component, the industrial gas turbine.

Early Developments

Designs for machines resembling the modern gas turbine, comprising a compressor, combustor and turbine, have existed for over a hundred years. Inventors such as Franz Stolze of Berlin had patents for designs which featured these components as early as 1873 (1). However, a practical gas turbine capable of producing any useful power was not built until 1909. This machine, designed by Armengaud and Lemale in France, was barely up to the job since it struggled to drive its own 25 stage compressor (2). Not surprisingly, their design failed to catch on.

It was another 30 years before the fundamental flaws in such prototypes were overcome. A lack of appropriate materials and an insufficient understanding of compressor aerodynamics hampered efforts both in Europe and the USA. Finally, in 1939, the machine which is widely acknowledged to be the first industrial gas turbine in commercial service was installed in Neuchatel, Switzerland by Brown Boveri.

The years before World War II brought about the major technical advances that many designers had been trying to achieve. This was not in the field of industrial gas turbines, but in a very similar technology, the aircraft jet engine. Intensive Government funded programmes were initiated in the UK and Germany as the two sides battled for superiority. Since the aircraft piston engines of the 1930s were reaching their technological limits, the first country to fit jet engines to their fighter aircraft would have a huge advantage in speed and power. The race was won by the Germans when Von Ohain's jet powered the first aircraft of its type in 1939. This was closely followed by the first flight using the Whittle engine in the UK in 1941 (3).

This wartime development led to the first major involvement of most of the current manufacturers of large industrial gas turbines. The company that manufactured the first jet engines in the UK and many

Opportunities and Advances in International Power Generation, 18–20th March 1996,
Conference Publication No. 419, © IEE, 1996

of its employees have since become part of GEC's gas turbine activities. GE of the USA was given a Whittle engine to clone and manufacture in the USA, starting in 1941. Many physical features of the Whittle engine such as the 'combustion cans' find an echo in current GE industrial designs, the first of which was built in 1948 (4). Meanwhile, Westinghouse was awarded a contract by the US Navy in late 1941 to design and build a completely new jet engine design. The programme gave rise to its first test flight in 1943 and its continuation led to Westinghouse's first industrial gas turbine in 1949 (5). In Germany, the post-war story was rather different. Siemens had 80% of its assets taken away, particularly from its Berlin facilities, by Soviet and American forces as part of the post war settlement (6). Siemens, along with all other German and Japanese companies, was also prevented from manufacturing combustion turbines until 1952. However, the company was able to recruit engineers such as the chief designer of Junkers the (Germany's jet engine manufacturer) to work on their first designs. The transfer of tacit knowledge from these wartime efforts was invaluable and helped Siemens to catch up with its competitors once the ban had been lifted. ASEA Brown Boveri (ABB), the inheritor of Brown Boveri's pioneering technology which appears to be independent of jet engine technology, have also benefited in many ways. For example, ASEA was contractor for the Swedish jet programme through its Stal-Laval subsidiary, starting in 1945 (7).

It is clear that the wartime programmes were very important in bringing gas turbines from experimental status to commercial reality, particularly in the UK and USA. However, manufacturers on the European mainland such as Siemens and ABB had initial designs that were much nearer to steam turbine technology. The turbine, compressor and combustor were essentially bolted together, with the large combustors (of which there were only one or two) resembling a conventional boiler. The integrated gas turbines produced by GE and Westinghouse also drew on steam turbine construction methods but were very different on the inside, with aero-engine derived blading and multiple combustors in an annular ring.

At the same time as the development of the industrial gas turbine, ideas about its use in combined cycles were not far behind. For example, the waste heat from GE's first industrial gas turbine was used elsewhere in the facility where it was installed. The concept of the combined cycle itself dates back much earlier to ideas by Emmet (8) for mercury steam plants. However such ideas were succeeded by the CCGT during the 1950s. Seippel and Bereuter (9) of Brown Boveri were among the first to do a substantial amount of work on this, which yielded the first installation in Luxembourg in 1956.

The Technology Matures

Although the idea of the CCGT was established, its initial applications were limited. The basic problem was that the gas turbine itself was still a novelty in the 1950s and unit sizes were not sufficient to compete with steam turbines which were going through a period of rapid development. The added complexity of a combined cycle meant that any CCGTs that were built were a rarity, and derived most of their power from the steam turbine portion. The gas turbine was used as a power enhancer rather than a prime source of generation. During the period to the mid 1960s, manufacturers of gas turbines were struggling to sell enough units despite the fact that it was possible to use them to power railway locomotives, ships and even tanks as well as electric generators. Many have admitted that consideration was given to pulling out of the business because there was no money to be made. However, the growth of the oil and gas industry in the USA, South America and the Middle East appears to have provided a growing source of income to keep things going.

The 1950s were a time for further experimentation. Due to the inferior efficiency of gas turbines, they were often fitted with all kinds of add-ons such as intercoolers to compensate. As is often the case, the greater complexity led to unreliability and increased suspicion about technological novelty. Experiments were also conducted in the UK and elsewhere to see how many different fuels could be burned in these machines. Initial trials held by GEC and others with coal gas were plagued by breakdowns and were abandoned.

The mid 1960s brought a significant breakthrough for the makers of gas turbines. Blackouts in the electricity systems of the UK and North America led to the installation of a large number of units for use in future emergencies. The ability of gas turbine to start up very quickly was a crucial factor in the decision to use it for this application. The resulting income allowed some manufacturers, particularly GE and Westinghouse, to introduce a lot of new technology from jet engines such as air cooled blades and high temperature materials to enhance power output and efficiency.

The new generation of machines that followed in the early 1970s were a lot larger, with unit sizes up towards 100MW, and more efficient. The heat from the turbine exhaust was now sufficient to raise enough steam to drive a steam turbine without supplementary firing. As a result, combined cycle efficiencies were beginning to exceed the 40% achieved by conventional steam plants. A co-incidental gas bubble in the USA helped get the first CCGT orders for GE and Westinghouse. Most were designed for mid-range service which was a much higher load factor than these larger gas turbines had been used to. Until then, the only machines that had worked for long periods without stopping had been small units in the oil and gas industry. A lot of these running together meant that if one failed it wasn't a major problem. Now, things were starting to change as unit sizes increased. The complexity of the new combined cycles, which demanded more sophisticated control and higher availability, forced manufacturers to look at the issue of reliability more closely. This was done with the active involvement of bodies such as the Electric Power Research Institute in the USA who served as a focus for the needs of utilities.

At approximately the same time, in 1973/4, the newly emerging gas turbine boom was strangled by the decision of the OPEC nations to quadruple the price of oil almost instantaneously. Harder times set in as oil and gas came to be viewed as too precious for use in power generation. This difficult period lasted until the mid-1980s, though most of the major companies managed to keep their operations going by virtue of their sheer size and diversity. Even Westinghouse, who closed down its domestic large gas turbine manufacturing facilities and transferred production to Mitsubishi (its Japanese partner) has now recovered.

The oil shocks of the 1970s were not entirely disastrous for the gas turbine since they encouraged some Governments to put money into research programmes that aimed to increase fuel efficiency. The Japanese Moonlight programme and the American High Temperature Turbine Technology programme are two examples which directly helped manufacturers to make further advances in relevant core technologies such as materials science. In the background, defence spending on jet engines was still very high. In the USA alone, $450m per year was spent between 1976 and 1986 (10).

The Market Recovery

The first signs of recovery for the gas turbine market outside the limited market in the oil and gas industry occurred in the USA with the introduction of the PURPA act in 1978. The legislation was designed to encourage the use of alternative fuels but was also a spur to the use of gas turbines in efficient combined cycle configurations, particularly in industrial applications. Once the US suppliers had begun to gain sufficient experience in the home market and address some of the reliability problems, they were in a position to start selling some CCGTs elsewhere with the help of their international affiliates. This initially happened in Japan and some developing countries such as Mexico and India via World Bank initiatives (11).

Despite the fact that some of these plants in developing countries suffered from poor performance, the gas

turbine boom had been restarted. In Europe, interest was patchy until the privatisation of the UK gas and electricity industry triggered the 'Dash for Gas'. Further encouragement was given by new environmental regulations and, to a lesser extent, the lifting of a 1975 EC directive restricting the use of gas for power generation. Of the main European equipment suppliers, ABB (formed in 1988 from the merger of ASEA and Brown Boveri) and Siemens had already begun to establish themselves as international suppliers of CCGTs during the 1980s. The newly formed GEC-Alsthom had discontinued its own work in large gas turbines due to a lack of support from the CEGB, and concentrated on building machines under license from GE of the USA. This arrangement also included some collaboration on gas turbine design, with GE retaining control of the essential core technologies.

By now, gas turbine unit sizes were passing the 200MW mark, with further application of new technology allowing combined cycle efficiencies to approach 55%. The sudden surge in worldwide demand meant that competition in the equipment industry has since become extremely fierce. This has accelerated new developments to the point where new models of gas turbine are now tested 'in the field' with the inevitable consequences for reliability. In order to compete with the American companies, particularly GE who have always had access to their own jet engine technology, the European manufacturers formed alliances with other jet engine makers. With technological developments from the 'steam turbine tradition' reaching a limit, Siemens have now begun to buy in technology from Pratt and Whitney, and ABB are using the experience of MTU (a German aero engine company) following a failed attempt to work with Rolls Royce. Rolls Royce have since formed a two way technology transfer agreement with Westinghouse who also have a tri-lateral alliance with Mitsubishi and Fiat Avio. Westinghouse have had to do this since they exited the jet engine business in 1960, and have increasingly had to rely on Mitsubishi during the 1970s and 1980s to stay in the gas turbine market. It is not clear how much the aero engine experience of Fiat has contributed to this alliance. It is hoped by all of these companies that their alliances will eventually lead to a CCGT of 60% efficiency, though GE is the only company to have released any details of an actual design so far. In any case, all concerned will have to address the problems that have caused the recent series of turbine failures before suspicious customers will invest in the next generation of technology.

To give some idea of the present status of CCGT installations worldwide, Table 1 shows the countries with the ten largest installed capacities (under construction and in operation) together with a summary of the reasons for high demand in each country.

TABLE 1 - CCGT Capacity Worldwide in Top Ten Countries (Source : SPRU CCGT Database)

Country	Confirmed CCGT Capacity (MW)	Reasons
USA	23000	PURPA Act, Repowering
Japan	14301	Diversity of fuel mix
UK	13215	Privatisation
Thailand	9070	Rapid Load Growth
India	7824	Rapid Load Growth
South Korea	7810	Rapid Load Growth
Indonesia	7243	Rapid Load Growth
Netherlands	5356	Tradition of CHP, liberalisation
Malaysia	5016	Rapid Load Growth
Taiwan	4817	Rapid Load Growth

The total capacity installed or under construction throughout the world has now exceeded 155GW.

ANALYSING THE SUCCESS OF THE CCGT

It now remains to give some indication as to the particular reasons for the apparent success of the CCGT with reference to the three themes mentioned earlier. The analysis presented here is not exhaustive but is designed to give some key points.

Technological Attributes

The gas turbine in both simple and combined cycle now has many advantages over other options, such as low capital cost, low emissions, speed of construction and fast start-up to cope with emergencies. However, when taking a longer term perspective to write a paper such as this, one further aspect seems to have been very important. This is the *flexibility* of generic gas turbine technology.

It is important to define what flexibility means in this context since it has many aspects. Some power plants are fuel flexible since they are able to burn a wide variety of materials, including 'difficult' fuels with a high water content. One example of such a plant would be the Fluidised Bed boiler. This technology has not been as successful as the CCGT. It has had some success on a small scale in the pulp and paper industry, and received a major boost from the USA's PURPA legislation which encouraged industrial cogenerators to burn unusual fuels. However, it has not yet become an option for the larger scale plants of over 300MW which still make up most capacity additions.

Another aspect of flexibility is siting. This is a strength of the gas turbine, particularly on a small scale. Since the 1960s, manufacturers have packaged such plants so that they may be delivered to a site and almost literally 'plugged in'. Diesel generators also display this characteristic but at smaller sizes. Larger gas turbines and CCGTs require less space than equivalent coal or nuclear stations and place less constraints on siting due

to lower environmental impact.

The aspect of flexibility that appears to have benefited the gas turbine so much is technological flexibility. The generic technology is key to two large industrial sectors - aircraft and power generation. It can also be applied to ships and used in pumping stations for pipelines. It could be said that nuclear power is similar since it is both a military and a civilian technology. However, it has always operated on a relatively large scale which precludes it from many of the applications that have used the gas turbine. It would be difficult to imagine nuclear powered oil rigs or oil pumping stations ! There is also the significant problem of waste disposal and decommissioning. As long as gas remains plentiful, the CCGT will always have an advantage from an environmental point of view unless the threat of Global Warming is seen to be a much higher priority.

The technological flexibility of the gas turbine has meant that it has been able to survive periods of low market demand in order to be used in a combined cycle. During the 1970s, the applications in the oil and gas industry kept demand for smaller machines above zero whilst the utility scale versions were not being sold. Also at this time, there was still a major investment being made in jet engines. Even if some of the major companies had pulled out of the industrial market, the technology would have continued to develop. GE in particular has always had the advantage of being in both jet engines and industrial gas turbines. This meant that core technological development was retained within the company.

Similarly, some have cited the early stages of the oil industry as the source of initial demand for small units during the late 1950s and early 1960s. When the first surge in demand occurred due to the blackouts of the 1960s, the gas turbine was ready to fulfil an unforeseen demand. Even though most of the emergency units installed were of the 'aeroderivative' type, with a modified jet engine driving a power turbine, the effect was to enable further development of industrial machines by the manufacturers. It is the nature of these manufacturers that will be examined next.

The CCGT Manufacturers

The current market for large industrial gas turbines and combined cycles is dominated by four networks of companies. Each network is linked together by a series of commercial and technological agreements which range from marketing accords to full blown technological collaboration. At the centre of these networks are four companies which have been amongst the most powerful suppliers of power station equipment throughout this century. They are GE and Westinghouse of the USA, Siemens of Germany and the Swiss-Swedish company, ABB. As mentioned earlier, all of these manufacturers have been involved in gas turbines and combined cycles since development started. Of course, there are many other companies that manufacture and sell CCGTs such as GEC-Alsthom and Mitsubishi. However, these other companies are reliant to some extent on the technology of the four main players. By maintaining control of key technologies and competencies, GE, Westinghouse, ABB and Siemens have been able to maintain their position as the market leaders. This has been further enabled by their size and diversified nature. At times when their gas turbine business was not making any money, they could afford to subsidise it or retain people with the necessary knowledge until the market improved. This may have been done by transferring staff to another division. For example GE may have moved resources between its aero engine and industrial divisions. Westinghouse certainly did until 1960 when it pulled out of the jet engine market. At this time, some of the staff from the division to be closed were transferred to the industrial business.

Siemens and ABB have not had the advantage of owning a significant jet engine division. However, the fact that their gas turbines have particularly close historic links with their previous steam turbine experience meant that this new technology was somehow familiar. It may be that a turbine-based technology such as the CCGT has become such a success because that is where the skills of the most powerful industrial players have been concentrated. A technology based on further development of traditional boilers such as the Pressurised Fluidised Bed demands different skills which may explain a relative lack of success. This is a tentative hypothesis but one that requires further investigation.

A final point about the equipment manufacturers relates to their attitude to new innovation. When the gas turbine and combined cycle were in the earlier phases of development, such attitudes were different to those of today. For example, Siemens was historically run by industrialists. These individuals were sympathetic to the wishes of scientists and engineers to experiment with new ideas, and would act as product champions. At the early stages of new technological developments, such a patient attitude is sometimes necessary to give ideas time to be fully evaluated. The gas turbine was no exception since it took approximately two decades to establish itself. Many in the industry are of the opinion that an embryonic gas turbine wouldn't get very much attention in the modern industry.

Government Involvement

The third and final factor that has helped the CCGT to become successful is direct and indirect support from Governments. This support has been continuous since the wartime development programmes for the first jet engines were initiated. The benefits of successive military projects in the UK and the USA have been considerable for the three main jet engine manufacturers, GE, Pratt and Whitney and Rolls Royce. It is now evident that the investment made in technology is helping the industrial gas turbine manufacturers to improve the performance of their products via technology transfer agreements or inter-business collaboration (in the case of GE). In the words of one industry observer, there is a whole supermarket of technology created by the jet engine programmes which is slowly being utilised by the power equipment companies. This supermarket contains everything from new alloys that can withstand higher temperatures to computer codes for advanced turbine blade profiles.

This transfer of expertise is not confined to recent years. As mentioned earlier, GE and Westinghouse first applied aero engine cooling techniques to their industrial turbines in the late 1960s. In addition, aero engines have been packaged as 'aero-derivatives' since the late 1940s. Such packages formed many of the emergency gas turbine units installed after the 1960s blackouts. Some of the modern advanced aeroderivatives currently on offer can be traced directly to military projects. For example, the LM6000 gas turbine from GE is a modified CF6 aero engine which was designed for a military transport plane. The aeroderivative version is one of the most efficient gas turbines around (rated at 40% in simple cycle).

More direct support of industrial gas turbines has been on a much smaller scale though there have been several programmes. Until the recent US initiatives by the Department of Energy, the most significant examples are the High Temperature Turbine Technology programme in the USA and the Japanese Moonlight project. Both were initiated in the 1970s to increase fuel efficiency. Westinghouse and GE were closely involved in the US programme and Mitsubishi was appointed project leader for the Japanese project.

In addition to direct funding of R&D, Governments and their associated institutions have aided the success of the CCGT in other ways, usually driven via wider political and economic objectives. The most striking case is that of the UK, where the decision to privatise the electricity and gas industries, together with the Conservatives desire to take revenge on the coal miners, has meant that almost all new investments in power plants have used CCGTs. Elsewhere, the World Bank has funded many of the first CCGTs in the Developing Countries, helping to establish the technology there. The rapid load growth seen in countries such as Thailand and Malaysia has meant a steady stream of CCGT orders. As liberalisation spreads to other areas of the world from its origin in the USA and Europe, the newly formed Independent Power Producers are using the CCGT on many projects in Asia and elsewhere due to its low capital cost and short construction time. This minimises the risks for IPPs since it allows them to recieve a very quick and profitable return on their investment.

CONCLUSIONS

Bearing in mind the amount of evidence to the contrary, the often repeated claim that the success of the CCGT is a triumph of the free market is plainly not true. This paper has shown that this success, if that is what it is, can only be explained by reference to a complex web of factors. Many of these have not been directly concerned with the CCGT at all ! Factors such as the desire of Governments to improve their military presence in the skies, or the lack of emergency provision in some power grids have had much more effect than direct R&D support.

This conclusion raises an important question. Since the success of the CCGT has little to do with direct public R&D funding, what are the implications for democratic decision making ? It seems that if the general public wants to develop or use a different technology such as wind turbines, Fluidised beds or gasification of alternative fuels, there is no mechanism for the expression of that wish. Instead, factors over which they have no control make such decisions. For the UK, the closure of a large portion of the coal industry in the face of widespread public opposition is a good example. If gas prices rise or supplies are unreliable and there is a need for coal again, a lot of flooded mines will not be of much use !

ACKNOWLEDGEMENTS

The research that led to this paper comprised many hours of interviews with relevant actors as well as a large amount of correspondence and literature reviews. It is not possible to acknowledge the source of all the information gained but I would particularly like to acknowledge the help of the following companies and institutions whose employees agreed to be interviewed. ABB Power Generation, Siemens KWU, GE Power Systems, Westinghouse Electric Corp., Mott Ewbank Preece, Rolls Royce Industrial Power Group, European Gas Turbines, Electric Power Research Institute, Strategic Power Systems Inc, OFFER, Montagu Smith and Company and London Electricity. I would also like to thank all of the members of SPRU Energy Programme, particularly my supervisors John Chesshire and Gordon MacKerron, who have collectively supported my work to date.

REFERENCES

1. Friedrich R, 1991, "Dokumente zur Erfindung der heutigen Gasturbine vor 118 Jahren", VGB Kraftwerkstechnik GmbH, Essen, Germany.

2. Constant EW, 1980, "The origins of the turbojet revolution", John Hopkins University Press, USA.

3. Scott P, "Birth of the jet engine", Mechanical Engineering, Jan 1995, 66-71.

4. Gorowitz B (ed.), 1981, "A century of progress - The General Electric Story 1876-1978".

5. Bannister RL, DeCorso D, Howard GS and Scalzo AJ, 1994, "Evolution of Heavy-Duty Power Generation and Industrial Combustion Turbines in the United States", Presented at the International Gas Turbine and Aeroengine Congress and Exposition, The Hague, Netherlands.

6. Owen Smith E, 1983, "The West German Economy", Croom Helm.

7. Jeffs E, "GT35 : The Aero Engine that Never Was", Turbomachinery International, May/June 1995

8. Emmet WLR, 1925, "The Emmet Mercury-Vapour Process", Trans of the ASME, 46, 253-285.

9. Seippel C and Bereuter R, 1960, "The theory of Combined Steam and Gas Turbine Installations", Brown Boveri Review, 47 783-799.

10. Williams RH and Larson ED, 1988, "Aeroderivative Turbines for Stationary Power", Annual Review of Energy, Annual Reviews.

11. Moore E and Crousillat E, 1991, "Prospects For Gas-Fuelled Combined-Cycle Power in the Developing Countries", World Bank, Washington, USA

COMBINED CYCLE GAS TURBINES AND CLEAN COAL

C H Buck

GEC ALSTHOM Power Generation Division, UK

INTRODUCTION

Coal is a fuel which cannot be ignored, its displacement by gas and oil as the primary feedstock in the generation of electricity is local and of uncertain endurance. Relative to gas and oil, coal has a significantly greater abundance and is far more evenly distributed around the world. Early development work targeted handling and basic combustion together with increasing plant size. The limited knowledge and low relative power consumption restricted the concern for the environment.

A widening of environmental awareness and the developing application of scientific study eventually generated studies into the consequences of the combustion of fossil fuels. These studies resulted in initial demands to limit the emission of acid rain inducing pollutants and, more recently, those products which promote global warming.

The concerns associated with pollution assisted the rarer but cleaner fuels, oil and (more particularly) natural gas to qualify as practical competitors.

Increasing power generation requirements, competition with other fuels, work in regions which were or are industrialised but rich only in coal have forced the production of alternative fuels from coal. Various processes have now been developed to enable the production of clean, relatively simple fuels from coal. This is Clean Coal Technology (C.C.T.)

Parallel development of power generation machinery and the gradual emergence of the gas turbine as an efficient, flexible and cost-effective machine in its own right provides us with the means to exploit a wide range of fuels. When combined with waste heat recovery, modern gas turbines are now accepted as viable base load generation plant.

The bringing together of the advancing clean coal technology and the high efficiency gas turbine to produce power reliably and cheaply with a low environmental impact is seen as inevitable and imminent.

When the gas turbine is integrated with clean coal technology, nations which are presently moving towards natural gas or distillate fuel to generate power cleanly will have an extra option. Nations which are presently burning coal in traditional equipment will have the means to convert to clean utilisation which is more likely to attract international support, and emerging consumers will be able to develop their own resources using clean and efficient plant.

Gas turbine manufacturers have been keenly aware of all the fuel options and have closely monitored the progress of clean coal technology. This effort was inspired by initial failures of direct coal combustion and studies into the effects of fuel contaminants on machine operation. Commercial and environmental factors have been developed to compliment the engineering considerations in maintaining an overview of future business

Various clean coal technologies have now advanced to a state where prototypes can be designed using the results of isolated component development.

European Gas Turbines has worked for a considerable period with research organisations and universities to investigate the applications and to develop various aspects of clean coal technology. As a result of this effort, EGT has chosen to be associated with British Coal in the application of the British Coal air blown gasifier and has developed a partnership with British Coal, Mitsui Babcock Energy and Stein Industrie to exploit the technology.

The complete plant is based on a gasifier which produces the gas fuel from coal, a fuel gas clean-up system and the gas turbine which burns the fuel. Since the gasifier is designed to convert only part of the coal feedstock a circulating fluidised bed is utilised to complete the burn out of the coal.

The system is referred to as the Air Blown Gasification Cycle (ABGC), and is related to the other members of the Integrated Gasification Combined Cycle (IGCC) family.

The EGT preference for the ABGC is based upon the perceived efficiency and the versatility derived from a wide ranging feedstock specification, achievable variations in steam conditions and flexibility of operation.

Opportunities and Advances in International Power Generation, 18–20th March 1996, Conference Publication No. 419, © IEE, 1996

FUEL

The central effort is focused on coal and related materials. The need to be able to use coal to produce power efficiently in an environmentally friendly manner is emphasised by a knowledge of its availability not just in the UK but around the world. Although estimates vary, it is generally accepted that at the present rate of production the coal reserves will last for 250 years whereas oil and gas will be exhausted in 60 years. It is significant that coal is widely distributed throughout the world whereas oil and gas are unevenly distributed and are more prone to politically influenced availability variations.

Coal is available as an indigenous fuel to many of the countries which will have to replace a significant capacity of obsolete plant and to developing countries which must now generate supplies to meet considerable increases in demand.

Environmental pressures which will result in new plant meeting stringent emission limits and competition with other fuels raise challenges which will have to be overcome if this abundant fuel is to be widely used.

The characteristics of coal vary over a wide range. Non carbon constituents can form a relatively high percentage of the make-up and can include significant quantities of polluting material as well as components which set up highly corrosive atmospheres at high temperatures.

The wide range of coal characteristics tends to produce a useful side-effect in that plant which is developed to run on coal might be modified to run on other feedstocks such as "Pet Coke" and "Sewerage Sludge" as well as Residual Fuel Oil, Shale Oil, Bitumen, Orimulsion, Biomass etc.

THE ABGC

As the basic development of clean coal technology and gas turbine technology have developed, it was inevitable that workers in one field would be aware of progress in the other and that the two would be considered as complementary. In bringing the clean coal technology and the gas turbine together, many variations of a fully integrated plant have been identified and considered.

The basic requirement is to manufacture from the coal feedstock a fuel which is acceptable to the gas turbine and which does not burden the exhaust with polluting components. Any solid

and liquid products of the process should be benign and preferably useful.

The use of a circulating fluidised bed under a reducing atmosphere to generate a fuel gas from the feed stock is common to the mainstream proposals. Workers adopt one of two basic methods - Air Blown or Oxygen Blown. An oxygen blown system gasifies the coal at atmospheric pressure to generate a medium (8 MJ/M³) calorific value gas and an air blown system gasifies the coal under pressure to generate a low (4 MJ/M³) calorific value gas.

After much development work and various other investigations, the partnership which includes European Gas Turbines has concentrated its efforts on the Air Blown Gasification Cycle. There are various factors which suggest that the ABGC has advantages over competing systems and these will become evident later.

The diagram shows schematically, the basic ABGC.

The coal is prepared to a suitable particle size range and is transported pneumatically by screw feeders through lock hoppers to the pressurised fluidised bed gasifier developed by the Coal Technology Development Division, British Coal. In parallel with the coal, a sorbent, normally limestone, is also delivered to the gasifier bed. The sorbent is added to isolate the sulphur component of the coal.

At the conical base of the gasifier is injected the main air supply in the form of a jet which penetrates a significant distance into the bed. This is the basis of the British Coal design and is referred to as a submerged spouted fluidised bed. Because of the powerful central jet, rapid and thorough mixing of the bed constituents occurs and uniform gasification is promoted due to the establishment of an even temperature. The mixing is further intensified by injecting more air through the secondary jets in the cone wall of the bed base.

In order to increase the control exercised over the process, varying quantities of steam can be added to the fluidising air at entry to the gasifier. By controlling the temperature due to partial combustion of the coal - to approximately 1000°C, the volatile constituents are driven off and the endothermic gasification reactions are completed.

The gasifier is not designed to completely convert the coal so that the size and cost of the gasifier can be controlled and the high overall efficiency of the plant can be maintained.

The solid coal based product of the gasifier (char) and the associated sorbent (with captured sulphur) are transported through pressure let-down lock hoppers to an atmospheric air circulating fluidised bed combustor - (CFBC) where combustion of the coal is completed. The heat generated in the CFBC is transferred to the steam system and the solid waste will be transported away for road base, block manufacture or other uses. The PFBC will be designed to accept coal as well as char to facilitate start up and increase flexibility. Obviously the size of the PFBC element can be increased to accept coal as a supplement to char to trim the size of the plant to the client's requirements. The gaseous exhaust from the PFBC will be cleaned conventionally before being emitted to the atmosphere via the gas turbine stack.

After leaving the gasifier the raw fuel gas is passed to a refractory lined cyclone separator where the majority of any solids carried over from the gasifier can be collected and passed to the CFBC. It is perhaps interesting to note that a cyclone having only a moderate efficiency is required to enable the ceramic candle filter downstream to operate efficiently. Before entering the ceramic candle filter the fuel gas is cooled to between 400°C and 600°C and heat transferred to the steam cycle.

In the interests of efficiency the fuel gas should be maintained at the highest possible temperature. However, metallurgical and design considerations related to the ceramic candle filter and gas turbine control valves downstream encourage the moderate degree of cooling.

The ceramic candle filter serves to polish the fuel gas of remaining solids. Multiple banks of the long ceramic filter elements become caked with the residual solids and a pulse-clean system will send intermittent pulses of clean fuel gas back into individual rows of elements to blow off the caked deposit. It has been found that admission and settling of some particles larger than an efficient upstream cyclone might allow past improves the efficiency of the ceramic filter as previously noted

The caked deposit discarded by the ceramic filter is char and is collected and transported to the CFBC.

Although not shown on the diagram, the fuel gas enters the primary control section of the plant where it can be bled off to vent through a flare stack and/or metered by control valves prior to admission in the gas turbine. Studies indicate that the flare stack will be required during the early stages of start-up and the later stages of shut down of the gasifier since the fuel gas composition cannot be controlled within reasonable limits for these short periods. Protection of the plant during rapid shedding of load from the gas turbine is also facilitated by the flare stack since thermal and chemical inertia of the gasification system has to be allowed for.

The main gas turbine fuel control system might be extended to allow the gas turbine to act as a calorimeter so that the control valve can be constantly re-ranged to permit a wider than usual range of calorific values to be accommodated. This will increase flexibility by allowing the load share between the gasifier and CFBC to be adjusted (thus possibly optimising the steam cycle). It will also permit the change over to different types of fuel.

The fuel will then be burnt in the combustion chambers of the gas turbine which will drive its own electrical generator. At this time development work proceeds to develop a combustion system which will burn the fuel gas and control the generation of oxides of nitrogen to the very low levels anticipated for future legislation. This work targets primarily the conversion of the small quantities of ammonia generated in the gasifier process.

To complete the cycle, the air required to operate the gasifier is extracted from the gas turbine immediately downstream of the compressor. Since this air must be passed through the gasifier, through the gas treatment and control systems and then back into the machine, it will have to be pumped by an auxiliary variable speed compressor. Careful selection and control of this auxiliary compressor will add further flexibility to the cycle.

The gasifier will have to be started after the gas turbine has been brought up to part load, in order to do this the gas turbine must be able to operate on a secondary, conventional fuel. It is intended that the gas turbine will be able to operate up to full load on the secondary fuel. This will obviously be particularly useful if the gasifier is required to be run at zero or part load.

The gas turbine is associated with a waste heat recovery unit (WHRU) contributing to the steam cycle.

It can be seen that because the gas turbine and the CFBC will both be able to operate independently, considerable latitude should be available for the definition and setting up of the overall control system. It is also possible that if a staged construction is required, revenue might be generated before all of the plant is commissioned.

PLANT OPERATION

Contributions to the overall plant outputs for a balanced plant would be approximately as follows: Gas Turbine 44%, GT WHRU 22%, CFBC 20%, other steam raising components (associated with the gasifier) 14%.

The gasifier and CFBC each would have a turn down ratio of approximately 3:1 permitting a similar overall integrated plant turn down. By changing over to conventional fuels the CFBC can generate between 7% and 20% of the total power in isolation and the gas turbine can turn down from 44% (total power) to zero.

Analysis of the dynamic performance of the ABGC indicates that the loading rates should be 3% minimum from 30 to 50% load and 5% minimum from 50 to 100% load. The work also indicates that step load increases above the ramp rate will meet the target of 5%. As might be appreciated, a useful load following capability will be available.

It will be noticed that flexibility is considered to be of paramount importance in our approach as existing attempts to demonstrate clean coal technology on a meaningful scale seem to have been very tightly integrated, giving rise to control and reliability problems. It should be noted that whilst we are positively designing in flexibility, the ABGC easily facilitates this practice due to the nature of the individual pieces of plant.

At this time we are carrying out a detailed study into the reliability and maintainability of the developing prototype plant design in order that the design may be as refined as possible at the prototype stage. It is also intended that the prototype plant will facilitate, by design, the use of 'add-on' features to study advanced clean-up and associated techniques.

EXPLOITATION STRATEGY

Studies show that commercial viability can be justified at sizes of integrated plant in excess of 400 MWe for typical UK applications. Prior to a primary effort to establish a 400 MWe plant, it is intended to establish confidence by building a 100 MWe prototype plant.

At this time, it is considered that a 100 MWe plant permits identification with a true 400 MWe commercial plant yet will not involve excessive extrapolation from available experience. A 100

MWe plant would involve a 30 Te/hr gasifier and a 40 MWe Frame 6B gas turbine.

EGT is working with its partners to identify a suitable site to build the prototype plant. As part of this effort, design work is underway to specify in detail all of the plant to enable accurate performance calculations to be made and to enable integrated plant layouts to be drawn.

In order to expedite the search for a suitable site for the prototype plant it is recognised that significant variations in scope are available. The options range from simple repowering applications, through intermediate systems to the stand alone, fully integrated package. A simple application might call for only the gasifier, gas clean-up and gas turbine to provide power and, for example, feed water heating. An intermediate system might provide power and steam. In both cases existing infrastructure for solids handling and power distribution might be employed.

In parallel with the effort to establish the prototype plant, work continues to extend the data and to refine the technology. This work includes developing improved emission control techniques, combustion systems and dynamic performance predictions.

CONCLUSION

A Prototype Integrated Plant to demonstrate viable Clean Coal Technology can now be designed and constructed with confidence. Identification of a host site for the Prototype Plant is an action being undertaken at this time. Work continues to advance the state of the technology which is being established to permit the design of a full 400 MWe and larger commercial plants.

Air Blown Gasification Cycle

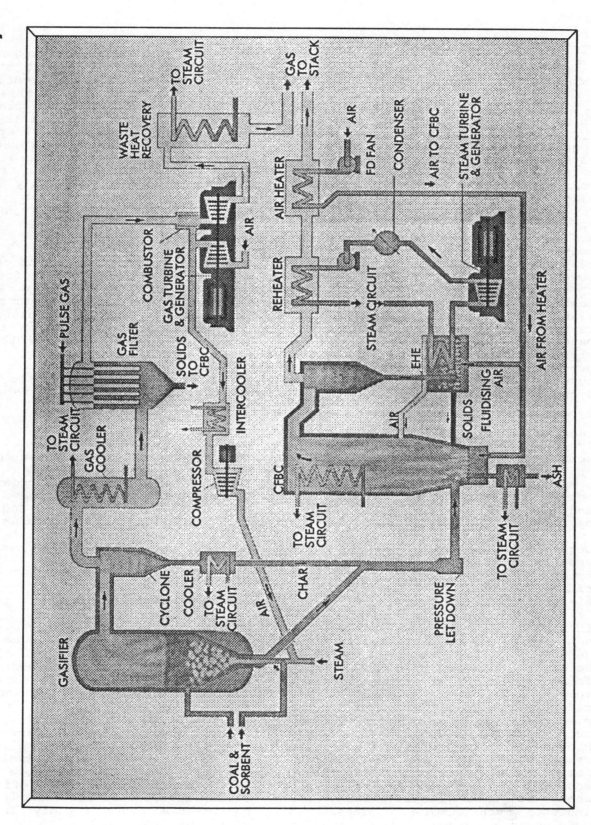

INDUSTRIAL COGENERATION IN THE UK

G Evans

National Power, UK

INTRODUCTION

In spite of its long history a certain mystery still surrounds the concept of cogeneration. Alternative names, such as combined heat and power, do not help and the recent emergence of combined cycle gas turbines (CCGT) also tends to confuse.

For this paper I will be consistent and adopt the term cogeneration. A cogeneration scheme is in most respects no different to any other power station. It is distinguished however by the fact that the waste heat from the electricity generating plant is harnessed and made use of rather than being thrown away. This waste heat could be in the form of steam to be used in an industrial process or perhaps hot water for space heating in a commercial building or district heating scheme.

In a conventional power station this heat is rejected to the environment. By capturing it in a cogeneration scheme fuel conversion efficiencies approaching 90% can be achieved. This compares with some 36% for a conventional coal-fired power station and some 52% for the CCGTs now in operation.

Cogeneration is not a new idea. It used to be very common for factories to have their own power stations which often supplied heat as well as electricity to the site.

In 1965 66% of the electricity consumed in the UK paper industry was generated on-site from cogeneration schemes. However, by 1990 this had fallen to 20%. The reasons for this very significant decline are essentially twofold. Firstly, the electricity grid system, established by the CEGB, reached maturity and proved able to supply electricity reliably and at lower real prices. Secondly, developments in boiler plant and the availability of relatively cheap oil made steam raising for industry much simpler. These two developments both acted against cogeneration schemes which gradually declined through the last two decades.

In recent years however cogeneration has been the focus of much attention and has had something of a renaissance. Several factors have made this happen but two are worthy of particular note. Firstly, in 1990 the electricity supply industry was privatised and competition was introduced to the generation of electricity. I will not review this here except to say that the new generating companies quickly saw the need to be more innovative in their approach to the market and rejected the CEGB's strong view that 'big is beautiful'.

Secondly, the previously held view that gas should not be used for power generation was relaxed by Government. This change had a number of drivers but the very high efficiency of cogeneration schemes and to a lesser extent CCGT schemes helped to change people's views.

These two factors acted together to renew interest in cogeneration and since 1989 some 1500MWe of new cogeneration capacity has been committed in the UK.

This paper briefly reviews the cogeneration market in the UK. It then focuses on industrial cogeneration schemes with reference to the schemes developed by National Power Cogen and a case study of one of them. It concludes by considering the future role for cogeneration in a competitive generation market.

THE COGENERATION MARKET

The cogeneration market can be broadly divided into three sectors. Firstly, there is the packaged cogeneration market generally supplying demands of less than 1MWe. These packaged units are almost always built around a reciprocating engine usually burning natural gas. They can be factory built and delivered to site as complete units making installation a relatively simple matter. Typical applications for this type of scheme are hotels, offices and large residential buildings. This market has grown rapidly in recent years and some 900 units are now in operation.

Opportunities and Advances in International Power Generation, 18–20th March 1996,
Conference Publication No. 419, © IEE, 1996

The second category is district heating. Here, a central power station will supply both heat and electricity to a discrete community of houses and offices. Though such schemes are quite common in some European countries their penetration in the UK has to date been limited. The primary reason for this is the very significant infrastructure costs of establishing a district heating grid. Such schemes have to compete with the established gas grid which offers 'point-of-use' heating to most buildings in the UK.

The third category is industrial cogeneration. This is where the most significant investment has occurred over the last five years. National Power established a separate division, National Power Cogen, to develop this market as it was clearly identified as having potential both in the UK and abroad.

INDUSTRIAL COGENERATION

There are essentially three criteria that have to be satisfied for a cogeneration scheme to be economically viable.

Firstly, fuel has to be available at a price competitive with that seen by the mainstream power generation market. Secondly, the specific capital cost of the scheme (£/kW) has to be as close to a conventional power project as possible. Lastly, the utilisation of the plant has to be as high as possible, ideally over 7500 hours/annum.

The traditional process industries - chemicals, petro-chemicals, paper and pharmaceuticals - prove to be ideal 'hosts' to cogeneration schemes. They operate on an effectively continuous basis, their heat and power demands often match well to cogeneration schemes and the size of the schemes can allow realistic specific capital costs to be achieved.

Unlike the smaller 'packaged' market, these schemes are designed specifically for a site. In contrast to traditional power projects their design tends to be 'steam led' with electricity almost taking the role of a by-product.

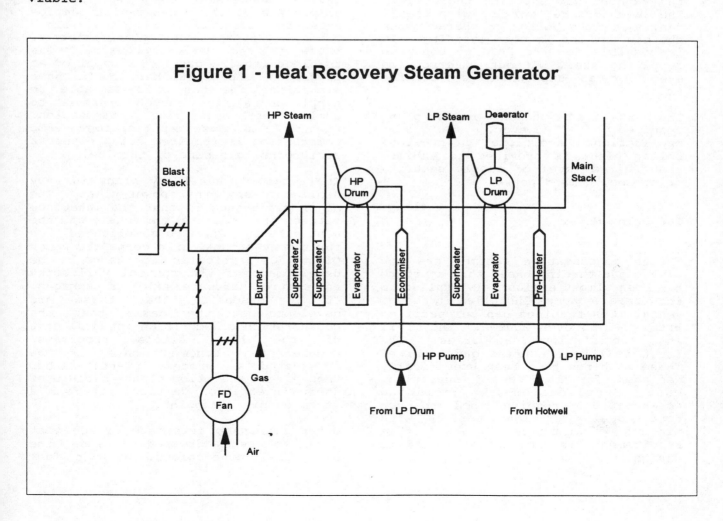

Figure 1 - Heat Recovery Steam Generator

SCHEME DESIGN

Steam Supply

The design of an industrial cogeneration scheme almost always starts from an analysis of the steam load to be met. Clearly, maximum and minimum demands are considered but just as importantly the rates of change of load and security of supply have to be considered. This analysis will help determine the type of boiler plant to be used, the number of independent sources of supply and therefore the level of redundancy.

The lead boiler or boilers in a gas turbine based cogeneration scheme will be the heat recovery steam generator (HRSG) that converts waste heat from the gas turbine into process steam. The schematic diagram (Figure 1) shows a typical HRSG. Though HRSGs are quite often un-fired, relying totally on the gas turbine's waste heat, this unit has additional burners in the inlet duct which provide supplementary and auxiliary firing. Supplementary firing acts with the gas turbine in operation raising the temperature of the exhaust gas entering the boiler. The same burners, working with the FD fan, allow the boiler to operate when the gas turbine is not operating. It is possible for the HRSG to continue supplying steam without interruption even for an unplanned gas turbine trip.

The HRSGs may be the only boiler plant required. However, in some applications, additional stand-alone boiler plant is provided to achieve the desired level of supply security or to meet demand peaks.

The Prime Mover

Though cogeneration schemes are not always gas turbine based this approach has been almost exclusively applied to industrial schemes installed in recent years. Gas turbines can be specified from 4MWe to over 200MWe. They fall into two distinct categories. The first is the industrial gas turbine. These machines have been specifically designed for land based operation. Many applications, including cogeneration, require them to run continuously for long periods with minimum maintenance and this requirement is reflected in their design.

The second is the aeroderivative. These engines are direct descendants of aero engines but packaged for fixed operation. Because of their design basis these engines are of lighter construction than industrial engines and have higher compression ratios. They tend to be more efficient in open cycle operation so that their exhaust gas temperatures are generally lower than their industrial cousins.

Both types of gas turbine are used in cogeneration applications. The choice is driven by both technical and economic factors. The higher efficiency of aeroderivatives is balanced by the lower maintenance costs of the industrial machines. Aeroderivatives also require gas to be supplied at higher pressures. If this is not available from the gas grid larger compressors are required increasing costs and reducing net electrical output. The design of the boiler plant also impacts here due to the difference in exhaust temperatures of the two types of machine.

Auxiliary Plant

Having addressed the two primary components of the cogeneration package attention transfers to the auxiliary plant. Perhaps most significant of these is the gas compressor. Gas turbines require gas to be supplied at pressures above 15 bar. In some situations the gas grid is able to supply at a high enough pressure to avoid the need for compression. However, in most applications some compression is required using either a reciprocating engine or turbine.

The remaining auxiliary plant is very similar to any other power plant. The major difference is that the interface between the cogeneration scheme and the host site is much more complicated than for a conventional grid connected power plant. Of particular importance is the development of the control philosophy and control hardware/software packages. It is essential that these are developed on the basis that the cogeneration scheme is an integral part of the host site's processes. Cogeneration schemes have to be considered as process plants rather than power generation plants because of their integration particularly from a steam supply viewpoint.

Other important interface issues that have to be addressed include the electrical connections, back-up fuel supplies and of course the environmental interface.

Taking the electrical connections first, full account has to be taken of the host site's electrical infrastructure and the connections to the grid system. Studies have to be carried out to ensure that system fault levels remain within switchgear ratings and that equipment thermal ratings are also adequate. Further studies will be required to ensure that the generator will remain stable for all credible fault conditions. Most importantly, agreement will have to be reached with the Regional Electricity Company for the connection of the generator and its mode of operation.

The supply of gas has already been discussed. Where the interruption of the gas supply cannot be tolerated back-up light oil supplies can be provided to the gas turbine many of which are capable of automatic changeover.

Last, but certainly not least, the environmental interface needs to be taken full account of. This requires the plant's emissions to be limited to levels set by HMIP and noise levels to be agreed with local planning authorities. For larger plants consent is required under Section 36 of the Electricity Act.

Typical Schemes

National Power Cogen has committed seven schemes to date. These cover the paper industry (Aylesford Newsprint and Jamont UK), chemicals (Albright & Wilson Oldbury and Whitehaven sites), pharmaceuticals (Sterling Organics) and petrochemicals (Lindsey Oil Refinery). The Aylesford Newsprint scheme is described here as a case study.

CHP AT AYLESFORD NEWSPRINT

Aylesford Newsprint Ltd (ANL) is a joint venture between SCA Graphic Paper and Mondi Europe producing 100% recycled newsprint. ANL has recently invested some £250m to construct a new paper machine, PM14, at its site at Aylesford, Kent. ANL is increasing its production capacity by 280,000 tonnes per annum (tpa) to 370,000 tpa. A total of 450,000 tpa of used newspapers and magazines will be

needed making an important contribution to national recycling targets. The paper making process is energy intensive requiring both electricity and steam. It is therefore an ideal application for cogeneration technology.

Background

Papermaking at Aylesford started in 1922 and provided the catalyst for other associated businesses to be developed at the site.

Steam has always been raised in a common utilities plant and distributed around the site to users. Electricity was also generated on-site but from 1984 until today has been imported from the grid.

Detailed studies confirmed that a cogeneration scheme could offer both technical and commercial benefits. ANL decided to move forward in the spring of 1992 and in June 1992 selected National Power Cogen to build, own and manage the operation of a cogeneration scheme at Aylesford. In November 1992 National Power Cogen and ANL signed a contract under which ANL would purchase heat and power from National Power Cogen's cogeneration scheme over an agreed contract term.

The cogeneration scheme was designed to initially supply the existing heat and power needs of the Aylesford site. These demands on average are some 35MW of electricity and 80 tonnes/hour of steam. This Phase 1 scheme (Figure 2) started operating in August 1994.

ANL's decision to build PM14 required the cogeneration scheme to be extended. Phase 1 had been designed to readily allow this extension and Phase 2 is planned to enter commercial operation at the end of 1995. It has increased the scheme's capacity to some 60MW electrical output and 200 tonnes/hour of steam. A simplified schematic diagram of the complete scheme is shown in Figure 3.

Design

The design brief for Phase 1 in summary was to provide a plant that would reliably supply the heat and power needs of the Aylesford site economically, either meeting or bettering current emissions standards.

Figure 2 - Phase I Cogeneration

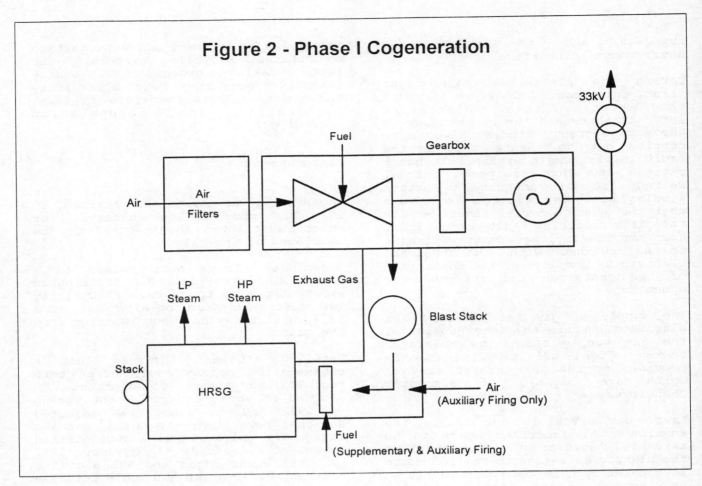

Figure 3 - Schematic Diagram

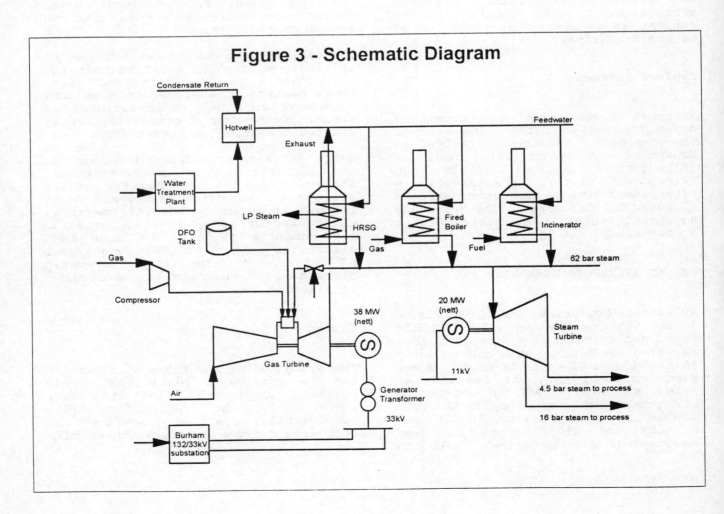

At the heart of Phase 1 is a GE Frame 6 gas turbine. The primary fuel for the gas turbine (GT) is natural gas. This is supplied to the site via a new pipeline installed by British Gas specifically for this scheme.

As the GT requires the gas to be supplied at a constant pressure (21 bar(a)) a gas compressor has been provided as part of the overall scheme. This compressor is a screw design being driven by an 11kV electric motor.

The compressed gas is delivered to the ten annular combustion cans of the GT. Should the gas supply be interrupted for any reason, a back-up supply of distillate fuel oil is brought on stream automatically. This fuel is stored on-site in a purpose-built storage facility.

The GT rotates at 5100 rpm. This in turn drives the 3000 rpm electrical alternator via a gearbox. The nominal electrical output of this package is 38MW.

The alternator generates at 11kV. This is transformed up to 33kV at which voltage it is supplied into ANL's distribution system.

The hot exhaust gases from the GT are ducted to the heat recovery steam generator (HRSG). This is a steam raising boiler designed to take heat input from two sources. Firstly, of course, it takes the exhaust gases from the GT. Additionally, however, gas burners are also provided to either top-up the heat input from the GT or, with the FD fan, fully replace it should the GT be unavailable. As the GT can also be run in 'open-cycle' (i.e. without the HRSG in service) using the blast stack the GT/HRSG provides an extremely flexible operating unit.

The HRSG is capable of producing some 80 tonnes/hour of steam from the GT exhaust gases alone. This increases to 120 tonnes/hour when the additional gas burners are brought into service.

The Phase 1 scheme also includes a new water treatment plant. This plant has two streams to achieve high reliability and capacity to allow further growth.

Under Phase 2, construction of which commenced late in 1994, an additional boiler has been added together with a back-pressure, pass-out steam turbine. With an eye to further expansion the new boiler has been designed to allow a second GT to be added in future. The boiler would then be converted to be a second HRSG of equivalent specification to the Phase 1 unit.

Construction

NP has, in line with EC procurement legislation, competitively tendered for both Phase 1 and Phase 2.

Construction of both phases has been complicated by the difficulties of working on a site with restricted access during the construction of the new paper machine. However, the site has remained operational throughout.

CONCLUSION

There is no doubt that cogeneration schemes offer a highly efficient way of converting primary fuel to useful energy. In the process industries cogeneration schemes are capable of offering both commercial and environmental benefits.

The environmental benefits have persuaded the Government to set challenging targets for the growth of cogeneration to help meet commitments made at the Rio Summit of 1992. The current target is for 5GW of cogeneration capacity by 2000.

Whether this target will be met depends on a number of inter-related factors. They include future gas and electricity prices, developments in electricity trading and the environmental pressures on larger energy users. However, in the UK's market driven energy market, unless specific legislation is introduced, the commercial benefits of cogeneration will be the driving force.

Acknowledgements

The author wishes to express his thanks to Aylesford Newsprint Ltd, the Combined Heat and Power Association and his colleagues who assisted him in the preparation of this paper, in particular Simon Jameson, Angela Hodgson and Paul Kavanagh.

a:ART-199/gce

POWER GENERATION USING HIGH EFFICIENCY AERODERIVATIVE GAS TURBINES

R P Smith

Parsons Power Generation Systems Ltd, United Kingdom

SYNOPSIS

The introduction of modern, high thrust aero-engines for aircraft propulsion has led to the development of a comparable range of new high power, high efficiency, aeroderivative gas turbines. This is typified by the Rolls-Royce TRENT, the world's most powerful aeroderivative and most efficient gas turbine of any type.

The development of these large, efficient aeroderivative GTs present specific opportunities for use in power generation and combined heat and power plant. This paper will outline the features that distinguish the TRENT aeroderivative GT and its merits in power generation. Consideration is given to the issues that differentiate the aeroderivative GT from industrial design GT plant and specific case studies are discussed which demonstrate the advantage of aeroderivative GTs.

INTRODUCTION

In 1994, Rolls-Royce announced the development of the TRENT aeroderivative gas turbine with a launch rating of 51 MW and 42 % efficiency, making it the largest aeroderivative and most efficient gas turbine of any type. Development of the TRENT gas turbine is now well advanced with the first engine currently undergoing testbed verification prior to delivery to the launch customer.

The TRENT aeroderivative GT is derived from the TRENT turbofan and powers the latest generation of large civil aircraft: the TRENT 700 entering commercial operation on the Airbus A330 while the TRENT 800 has been developed for the Boeing 777. The TRENT family of engines are themselves developed from the proven RB211 introduced in 1972, of which nearly 300 are currently in industrial service for power generation and gas pipeline pumping service.

TRENT AERODERIVATIVE GT

Design Features

The TRENT GT has been extensively described, Buxton and Toman (1). It comprises the established three shaft design of the RB211 which provides enhanced stage loading, aerodynamically optimised rotational speeds and compact engine design leading to tighter tip clearance control and hence reduced performance degradation. The industrial version of the engine retains a high degree of commonality with its aero parent as shown in Figure 1.

The Low Pressure (LP) shaft consists of a two stage compressor and five stage turbine and is directly coupled to the generator. Adjustment of the LP compressor blade angle enables the LP shaft to run at 3000 rpm or 3600 rpm to suit both 50 Hz and 60 Hz operation. The free running Intermediate Pressure shaft and High Pressure shaft have an eight stage and six stage compressor respectively each driven by a single stage turbine. The overall pressure ratio through the compressor is 35 : 1.

The combustion system consists of eight radial combustor pots capable of providing dual fuel firing. Operation on Natural Gas uses the Dry Low Emissions (DLE) combustor based on the, pre-mix, series staging, lean burn principle developed for the RB211. This was the first aeroderivative GT to enter commercial operation using a DLE system. The system enables full operational flexibility over a wide load range while maintaining low NOx, CO and UHC levels. Operation on liquid fuel utilises water injection for NOx control.

The TRENT GT is packaged into a self contained power generation plant by Westinghouse Electric Corporation and includes all the necessary components to enable simple cycle operation. The packaged plant, known as the TRENT EconoPac, provides a compact arrangement suited to minimising land utilisation as shown on Figure 2.

Performance

Basic plant performance in simple cycle and combined cycle gas turbine (CCGT) operation is shown in Table 1.

Opportunities and Advances in International Power Generation, 18–20th March 1996,
Conference Publication No. 419, © IEE, 1996

TABLE 1 - TRENT GT Performance

Simple Cycle [1]

Gross Power	51 190	kW
Gross Heat Rate (LHV)	8660	kJ/kWh
LP Turbine Speed	3000 / 3600	rev/min
Compressor Ratio	35:1	
Turbine Inlet Temperature (est)	1232	°C
Exhaust Mass Flow	159.6	kg/s
Exhaust Temperature	425	°C

Combined Cycle [2]

Nett Power	62709	kW
Nett Heat Rate (LHV)	6970	kJ/kWh

1 *Gross performance at ISO conditions, natural gas fuel, with turbine auxiliary loads but excluding inlet and outlet losses.*
2 *Nett performance at ISO conditions, natural gas fuel, 100 / 250 mmWG inlet/outlet losses, 45 mmHgA back pressure.*

With a gross power in excess of 50 MW, the TRENT GT is some 25 % larger than any other aeroderivative GT currently available. Similarly the gross heat rate of 8660 kJ/kWh (equivalent to 42% efficiency) is some 2 percentage points better than other modern aeroderivative units and the most advanced 'G' class industrial design GTs.

Advantages of the TRENT Aeroderivative GT

While the aeroderivative TRENT GT has been specifically developed for application in power generation, the commonality with its aero parent leads to a number of beneficial advantages in the industrial market and include:

O High efficiency
O High availability and reliability
O Small footprint area
O Ease of transport and installation for remote sites
O Maximum factory assembly and testing leading to short installation times
O Modular build
O Low maintenance requirements
O Low starting power requirements
O Rapid starting from cold
O Low thermal mass leading to short cool down periods
O GT removal / exchange within 36 hours

SIMPLE CYCLE APPLICATIONS

The efficiency advantage of the TRENT GT leads naturally to its consideration for simple cycle application. The two principal considerations for simple cycle operation are the required power output and utilisation factor.

Economic studies have shown that for mid-merit order plant (ie approx 4000 hours operation / year) with a nominal rating of 50 - 100 MW the TRENT GT offers a competitive cost of electricity due to the combined benefits of lower fuel consumption and lower maintenance downtime.

The advantages of large aeroderivative GT plant have also been considered for peak lopping and generation reserve application. A recent techno-economic study for a 200 MW+ oil fired GT installation considered the viability of a multiple TRENT installation versus:

i   a multiple industrial GT installation and
ii a single large industrial GT installation, each of equivalent rating.

For this study the capacity factor was only some 350 hours / year at approximately 75 MW. However, the plant had to also be capable of providing 200 MW+ over a 900 hour period during the summer (high load) months as cover for a main steam turbine unit trip on the electrical system. The results of the economic assessment are shown in Table 2.

TABLE 2 - TRENT GT Simple Cycle Economic Comparison

Option	Discounted Cost US$/kW	Electricity Price US¢/kW
4 TRENT	614	36.7
4 Industrial GT (equivalent rating)	760	42.3
1 Industrial GT (equivalent rating)	672	41.6

The results show that a multiple TRENT installation offers the most economic option. A number of factors contributed to this competitive advantage, all resulting from the unique design features inherent in a large aeroderivative GT. The two principal factors are:

○ The high efficiency results in a lower distillate fuel oil burn, which being a relatively expensive fuel leads to a significant reduction in operating cost. In the case of the single industrial unit this is particularly noticeable due to the relatively low load level of 75 MW resulting in a part load operating efficiency of approximately 20 % compared to the need to run only two TRENT GTs at close to full efficiency. This significantly offsets the higher capital cost of a multiple unit.

In the case of the multiple industrial GT option, while it is also possible to run only two units there is still a 4-5 percentage point efficiency advantage. The differential capital cost between a multiple aeroderivative and industrial GT is, however, also much smaller.

○ The faster start-up capability of an aeroderivative GT enables it to provide rapid reserve response. The respective normal start-up times for an aeroderivative and industrial GT are approximately 10 and 30 minutes from standstill to full load. Out of this period approximately two-thirds of the time is taken in running up from standstill to synchronous idle while the remaining third is used to load the machine from synchronous idle to full power.

It is therefore evident that the TRENT GT can provide full power reserve capability in 10 minutes from standstill whereas to provide the same power within 10 minutes from the industrial units it is necessary to run the GTs at approximately 5-10 % load. Over the 900 hours reserve requirement this results in a very high operating cost due to the low part load efficiency, whereas the TRENT GT provides reserve capability at zero operating cost.

Technically the multiple TRENT option was equal to, or better than, the industrial alternatives due to the advantages already stated and in particular:

○ Higher availability and reliability through quick engine and module exchange.
○ Lighter ground loadings leading to reduced foundation requirements.
○ Short installation times - minimal site disruption.

COMBINED HEAT AND POWER APPLICATIONS

Combined heat and power applications can fall into one of two categories:

i GT plus heat recovery steam generator (HRSG)
ii Combined cycle with process steam extraction (this latter case is covered in the next section).

Generally applications requiring process heat need either hot water or low grade steam, typically in the range 2 - 15 bar, saturated or slightly superheated. Aeroderivative GTs are ideally suited to these applications due to the matching of their relatively low exhaust gas conditions to the process steam conditions.

The process steam raising capability of the TRENT GT is shown in Figure 3. The optimisation of combined heat and power plant is normally dictated by the process heat requirement, any excess or shortfall in the electrical connection can be accommodated in the connection to the local utility. Where necessary, supplementary firing can be incorporated to increase the steam raising capability. For maximum benefit a combined heat and power plant using only a GT + HRSG needs to have a reasonably steady heat load so that the GT does not have to be operated at part load for lengthy periods (or, alternatively, operated with significant steam dumping). For applications where the process steam is subject to wide variations, the combined cycle with process steam extraction is the logical choice.

COMBINED CYCLE APPLICATIONS

The combined cycle performance for the TRENT GT is shown on Table 1. With a heat rate of 6970 kJ/kWh (equivalent to 51.6 %) the TRENT offers one of the most competitive CCGT efficiencies in its class.

In terms of steam cycle performance, the relatively low exhaust temperature of an aeroderivative GT coupled with the low steam turbine power outputs tends to lead towards the adoption of a 2½ pressure, non-reheat cycle design. The 2½ pressure cycle has high and intermediate pressure (HP and IP) circuits to supply steam to the steam turbine. A low pressure (LP) circuit is also included but only for deaerating duty. The adoption of the 2½ pressure, non-reheat cycle is considered to represent the most cost effective option in terms of first and through life costs. However, where capital cost is less important than efficiency, the use of a three pressure, no-reheat cycle is worth considering. Similarly a single pressure cycle may be appropriate if capital cost is a dominant factor.

Simplicity in the design of the steam cycle has been a key feature to ensure maximum reliability and availability. Similarly HP and IP pressures have been selected to ensure the best compromise between performance and flexibility to meet varying process steam requirements without significant re-engineering (refer to 'TRENT Reference Design').

The use of a pure CCGT plant with no process steam in the 60 - 120 MW class (equivalent to a one or two GT installation) is well suited to developing networks or distributed power applications.

Application to CHP

Most aeroderivative applications are generally centred on a requirement for heat and power and therefore tend to be industrial power plant within a process plant. One such application has been investigated for application within a UK manufacturing plant.

The average heat and power demands for the plant are:

- Minimum process steam flow   2.6 kg/s
- Maximum process steam flow   11.3 kg/s
- Annual steam consumption   150.0 kTe

- Minimum electrical demand   5.0 MW
- Maximum electrical demand   20.0 MW
- Annual Energy consumption   85.0 GWh

As can be seen, this application has a 4.5 : 1 variation in process steam flow and, as can often happen with such plant, the requirement for maximum heat and power can occur at the same time. The need for flexibility in the operation of the plant is thus a necessity and the use of a steam turbine generator provides a useful sink for surplus steam and thus generates valuable MW. Surplus power can be readily exported to the local utility grid and, because the plant rating is below the 100 MW threshold, the unit does not have to be load dispatched, thereby enabling any amount of power to be exported at any time.

The basic steam cycle design is shown in Figure 4. Process steam is drawn off the IP circuit which has been set at the relatively high pressure of 15 bara to meet the process requirements. This is somewhat higher than the optimum pressure for a pure CCGT plant (typically 5 - 6 bara) but provides a simpler plant arrangement and the slight loss in electrical output is compensated by the lower plant cost. During periods of low steam demand any excess steam in the IP circuit is inducted into the steam turbine to provide additional electrical power. At periods of high process demand additional steam is extracted through the same IP connection having first done useful work in the HP section of the steam turbine.

Economic modelling of the plant has shown the TRENT CCGT plant to be economically viable. An independent economic model compared the TRENT CCGT with a close competitor and showed a clear performance advantage to the TRENT with a two percentage point better internal rate of return (compare 13.7 % for the TRENT versus 11.4 % for the competitor). For the same financing arrangements the improvement in internal rate of return leads to 10 % improvement in return on equity and lower debt coverage ratios.

Economic analysis also shows that the TRENT CCGT is not particularly sensitive to realistic changes in output, availability and O&M costs.

TRENT Reference Design

The thermal cycle design for the above application is based on current development of a TRENT CCGT / CHP Reference Design. The concept of the Reference Design has been discussed by Johnson and Smith (2) and has several demonstrable advantages:

- Rapid, accurate and competitive responses to requests for proposals
- Established base specification from which project specific design can be developed
- Reduction in project programme
- Reduced duplication of engineering and procurement
- Increased quality

The Reference Design incorporates a high degree of modularization to maximise the advantages of replication in design engineering while reducing the site or customer specific engineering and installation time and costs. The modular design allows the major components to be located in fixed relationships to each other, thus providing a firm basis for establishing bulk material quantities and construction schedules, yet retains flexibility to meet a customer's unique requirements through the provision of a number of standard options. Experience with the Reference Design concept has shown that some 60 % - 75 % of a typical project can be replicated with only some 25 % requiring major new design effort to satisfy a Client's specific needs. In addition, the high degree of detail information available in the early stages of a project enables balance of plant systems to be rapidly specified. The Reference Design therefore allows early and rapid resolution of key issues and thus enables financial institutions to gain an early confidence in the financial viability of a given project. As the role of Independent Power Producers expands throughout the world, this feature will become ever more important in the development of power projects.

CONCLUSIONS

Aeroderivative GTs offer several advantages over their industrial design cousins, in technology, project implementation and maintenance. With regard to performance, modern aeroderivative GTs offer a very efficient form of simple cycle energy conversion and this is epitomised in the TRENT GT. The high efficiency of the TRENT makes it an obvious choice for simple cycle application, not only in mid-merit order duty but also for peak lopping and spinning reserve. Economic studies have shown that the high efficiency more than compensates for the higher first cost that can be expected with a state-of-the-art technology development from a modern aero-engine.

Modern aeroderivatives are also a viable choice for CCGT and CHP applications where power in the 60 - 120 MW class is required. With a combined cycle efficiency in the order of 51½ % the TRENT GT is one of the most competitive options available in its class and offers a financially sound business case.

Modern aeroderivative GT plant are a well accepted option for energy generation and this trend is expected to continue with the TRENT and future developments.

ACKNOWLEDGEMENTS

The Author wishes to thank Parsons Power Generation Systems Ltd for their permission to publish this paper. Thanks are also extended to my colleagues in Parsons Power Generation Systems Ltd, Rolls-Royce Power Ventures Ltd and Rolls-Royce Industrial & Marine Gas Turbines Ltd for their assistance in preparing this paper.

REFERENCES

1 Buxton RD and Toman R, 1994, "TRENT ECONOPAC DEVELOPMENT INITIATED", CEPSI 10, Vol 4, Pages 934-939

2 Johnson KW and Smith RP, 1994, "50 Hz COMBINED CYCLE REFERENCE PLANT DESIGNS", Power-Gen '94, Vol I, Pages 765-778

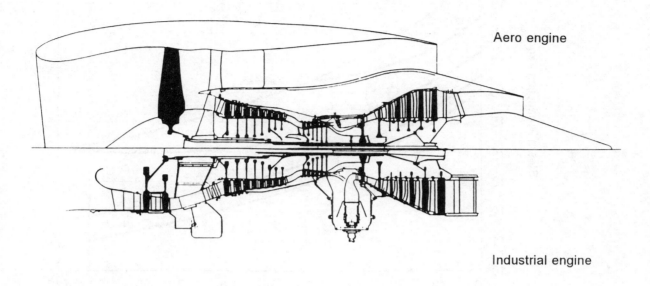

Aero engine

Industrial engine

Figure 1: TRENT Aero versus Industrial Comparison

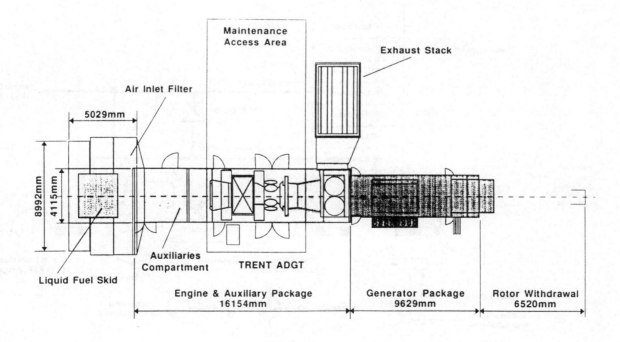

Maintenance
Access Area

Exhaust Stack

Air Inlet Filter

5029mm

8992mm

4115mm

Auxiliaries
Compartment

TRENT ADGT

Liquid Fuel Skid

Engine & Auxiliary Package
16154mm

Generator Package
9629mm

Rotor Withdrawal
6520mm

Figure 2: TRENT EconoPac

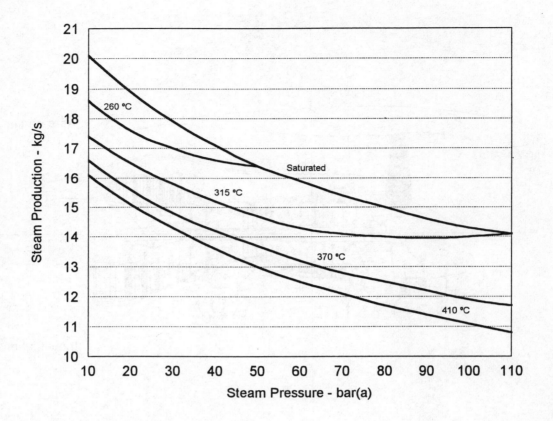

Figure 3: TRENT Steam Raising Capability

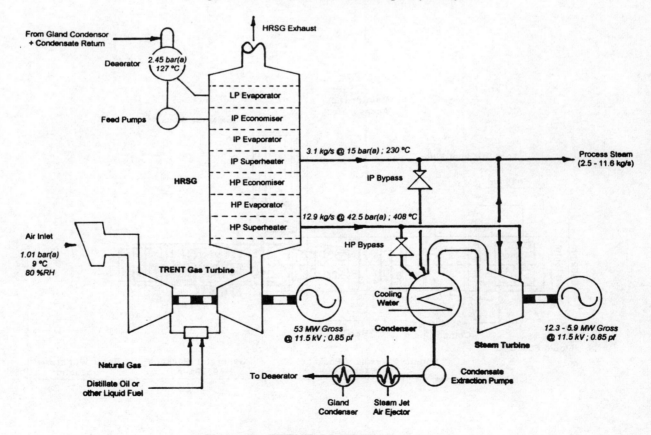

Figure 4: TRENT CCGT Thermal Cycle

A CONSTANT FREQUENCY CONSTANT VOLTAGE VARIABLE SPEED STAND ALONE WOUND ROTOR INDUCTION GENERATOR

R.S. Pena   G.M. Asher   J.C. Clare   R. Cardenas

The Department of Electrical&Electronic Engineering, University of Nottingham, Nottingham, England.

1. INTRODUCTION

Remote villages and communities isolated from the utility power have to generate their own electric power. When the prime mover of the generator has variable speed, as is the case when wind power is used, the electric generator must be able to generate constant voltage and frequency at the terminals regardless of the shaft speed. A doubly-fed induction generator (DFIG) may be used in this situation because it can generate constant voltage and frequency at variable speed. The DFIG has also the characteristic of generating power from both the stator and the rotor when it operates at supersynchronous speed. This characteristic can only be exploited if bidirectional power flow capability for the rotor excitation is provided.

It is well known that a DFIG can supply power to an isolated load at constant frequency by imposing rotor currents with slip frequency , Vicatos and Tegopoulos (1). The stator voltage can be controlled by regulating the amplitude of the rotor excitation current, Jeong and Park (2) and Bogalecka (3). Indirect control over the stator voltage can be achieved by regulating the flux in the machine.

In this paper an experimental investigation into using a DFIG to supply an isolated load is presented. An schematic representation of the system is shown in Figure 1. The system uses a Scherbius scheme with two voltage-fed current-regulated PWM converters, namely the rotor converter and the front-end converter. This allows bidirectional power flow between the stator and the rotor through the DC link and hence operation at sub and supersynchronous speed with low distortion currents is achieved. The stator voltage is indirectly regulated by controlling the stator flux magnitude. The stator frequency is kept constant as the rotor speed varies by imposing slip-frequency rotor currents in the machine.

The machine control strategy uses a vector control scheme based on d-q coordinate transformations. The reference frame is oriented along the stator flux vector position, Leonhard (4). The d-axis rotor current is used to control the stator flux and the q-axis rotor current is used to force the correct orientation of the reference frame.

The front-end converter is also controlled using vector control techniques, Jones and Jones (5). A d-q reference frame is used and it is oriented along the stator voltage vector position. This allows and independent control of the active and reactive power flowing between the generator terminals and the front-end converter.

The experimental rig consists on a 7.5 KW, six pole induction machine driven by a DC machine and two 5KW bipolar transistor inverters. The control strategies for the machine and the front-end converter have been implemented on a network of T800 transputer processors. The operation of the generator has been verified for a wide range of rotational speed namely from 500 rpm to 1500 rpm (slip=0.5 to slip=-0.5). Several experimental results showing the operation of the system for a wide range of steady state and transient conditions are presented..

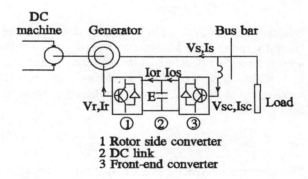

Figure 1. Stand-alone wound rotor generator

2. INDUCTION GENERATOR CONTROL

For variable speed operation the DFIG must supply constant voltage and frequency to the isolated load. Assuming that the voltage drop in the stator resistance is negligible, the stator voltage can be regulated indirectly by controlling the stator flux magnitude. Equation (1) describes the DFIG in a synchronous rotating d-q reference frame, Leonhard [4], where w_e is the rotational speed of the frame, i_{ms} is the stator magnetizing current and $w_{slip}=w_e-w_r$ is the slip frequency. v_{ds}, v_{qs}, v_{dr}, v_{qr}, i_{ds}, i_{qs}, i_{dr}, i_{qr}, λ_{ds}, λ_{qs}, λ_{dr}, λ_{qr} are the stator and rotor voltages currents and fluxes in the d-q frame. L_s, L_o, L_r are the stator, magnetizing and rotor

Opportunities and Advances in International Power Generation, 18–20th March 1996, Conference Publication No. 419, © IEE, 1996

inductances respectively. R_s and R_r are the stator and rotor resistance respectively.

$$\lambda_{ds} = L_s i_{ds} + L_o i_{dr} = L_o i_{ms}$$
$$\lambda_{qs} = L_s i_{qs} + L_o i_{qr} = 0$$
$$\lambda_{dr} = L_o i_{ds} + L_r i_{dr}$$
$$\lambda_{qr} = L_o i_{qs} + L_r i_{qr}$$
$$v_{ds} = R_s i_{ds} + \frac{d\lambda_{ds}}{dt} \qquad (1)$$
$$v_{qs} = R_s i_{qs} + w_e \lambda_{ds}$$
$$v_{dr} = R_r i_{dr} + \frac{d\lambda_{dr}}{dt} - w_{slip} \lambda_{qr}$$
$$v_{qr} = R_r i_{qr} + \frac{d\lambda_{qr}}{dt} + w_{slip} \lambda_r$$

From (1) using the condition of stator flux orientation, $\lambda_{qs}=0$, the following constrain is obtained:

$$i_{qr} = -\frac{L_s}{L_o} i_{qs} \qquad (2)$$

Considering the definition of i_{ms} given in (1) together with the constrain in (2) the following expression can be obtained, Leonhard (4):

$$T_s \frac{di_{ms}}{dt} + i_{ms} = i_{dr} + \frac{1+\sigma_s}{R_s} v_{ds}$$
$$T_s * i_{ms} * w_e = i_{qr} + \frac{1+\sigma_s}{R_s} v_{qs} \qquad (3)$$

Equation 3 shows that i_{ms} can be controlled using i_{dr}. The current i_{qr} constitutes a degree of freedom and can be used to force the orientation of the reference frame along the stator flux vector. Considering the second relationship in (3) the following expression is obtained for i_{qr}:

$$i_{qr} = T_s * i_{ms} * w_e - \frac{1+\sigma_s}{R_s} v_{qr} \qquad (4)$$

Therefore the reference current i_{qr} that forces the q-axis stator flux component to zero can be obtained from either (2) or (4). In the practical implementation of the system, the restriction given by (2) has been preferred to force the orientation mainly due to considerations of noise in the measurements. Under these conditions a decoupled control of the stator magnetizing current is obtained. The two current i_{dr} and i_{qr} are then imposed in the rotor using a voltage-fed current-regulated inverter. Constant stator frequency is achieved by applying rotor currents with a frequency equal to the slip frequency.

$$w_{slip} = w_e - w_r$$
$$\theta_{slip} = \theta_e - \theta_r \qquad (5)$$

θ_e is the stator flux vector position and it is obtained by integrating the stator frequency w_e, which is set in software. The rotor position θ_r is measured with a 720 pulses per revolution encoder.

3. FRONT-END CONVERTER CONTROL

The front-end converter must maintain the DC link voltage constant when power in the DFIG is flowing into the rotor (subsynchronous operation) and when power flows out of the rotor (supersynchronous operation). The control strategy uses a d-q frame aligned with the stator voltage vector position.

The expression for the active and reactive power in a synchronously rotating d-q reference frame is given by:

$$P = v_d i_d + v_q i_q \qquad Q = v_d i_q - v_q i_d \qquad (6)$$

With the orientation of the reference frame as stated above, the q-axis voltage $v_q=0$ and the active and reactive power can be controlled using i_d and i_q respectively.

The error between the DC link voltage and the reference is processed by a PI controller which sets the reference value of i_d. The reactive power can be adjusted using i_q but it is normally set to zero to give unity displacement factor operation. The i_d and i_q currents are also controlled using a voltage-fed current-regulated inverter.

The voltage vector position angle, θ, needed for the orientation of the frame, can be obtained either from the stator voltage measurements or from the stator flux vector angle, θ_e. The stator flux vector position does not have noise, because it is generated by software, and has been preferred in the practical implementation. The voltage vector angle is therefore given by:

$$\theta = \theta_e + \pi/2 \qquad (7)$$

4. IMPLEMENTATION

The overall control of the system has been implemented using a network of transputer processors, Asher and Sumner (6), as shown in Figure 2.

Transputer T1 is used as an interface between the transputer network and a PC-clone which allows the user to interact with the system to set operating points or to capture and graphically display on-line data (speed, currents, voltages, power, etc). Transputer T2 performs data acquisition of the stator and rotor currents and the stator voltages. It also implements the control loops for the magnetizing and the d-q axis rotor currents and generates the PWM demand voltages for the rotor converter.

Transputer T3 in conjunction with a timing board generates the PWM pattern for the rotor converter using a 1KHz carrier regular asymmetric sampling PWM scheme. Transputer T4 performs the control of the front-end converter, including The DC link voltage control, the d-q axis current control and generation of the PWM pattern. Again a 1KHz regular asymmetric sampling PWM scheme is used.

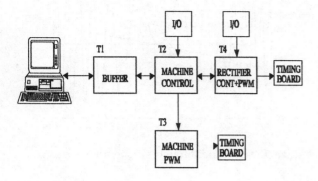

Figure 2. Transputer network

5. EXPERIMENTAL RESULTS

The overall system control has been tested for operation below and above synchronous speed. A DC drive was used to regulate the speed of the machine at different settings. The magnetizing reference current is set to 6.0A, which corresponds to approximately 220V at the stator terminals. The DC link reference voltage is regulated at 550V.

5.1 Front-end converter experimental results.

Figure 3 shows the DC-link voltage E, the phase voltage, V_s, and the front-end converter current, I_a, waveforms in the rectifier mode, when the machine is operating at subsynchronous speed.

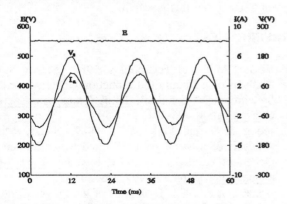

Figure 3. Front-end experimental results. Rectifier mode

Active power is injected into the rotor through the DC link . Since the reference i_q is set to zero, the phase displacement between the front-end converter current and the phase voltage is nearly zero. This demonstrates that the influence of the stator resistance is negligible when using the stator flux vector position to calculate the stator voltage vector position.

The operation of the front-end converter in the inverter mode, for supersynchronous operation of the DFIG, is shown in Figure 4. In this case the phase displacement between the voltage and the current in the front-end converter is 180° because power is flowing out of the rotor and it is supplied to the load through the DC link.

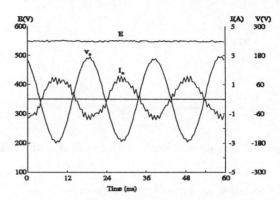

Figure 4. Front-end experimental results. Inverter mode

5.2 Load impact.

In this test the generator is initially operating at 700 rpm and supplying a resistive load of 550 W. A step increase of 1100 W is applied in stator terminals. Figure 5 shows the response to a step increase in load of i_{ms} and the waveform of the stator voltage. There is a drop in i_{ms} of approximately 12%, which is also noted in the voltage. However the control is able to recover magnitude of the magnetizing current and hence the stator voltage in two cycles.

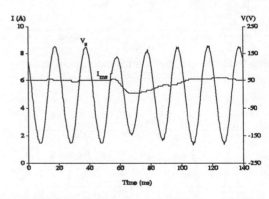

Figure 5. Magnetizing current and stator voltage

5.3 Inductive load operation.

The steady state performance of the DFIG supplying an inductive load at supersynchronous speed, s=-0.3, is shown in Figure 6. The inductive load was emulated by forcing the front-end converter to operate with a lagging phase displacement. By setting a negative i_q reference current in the front-end converter an inductive load is "seen" from the machine terminals. The Figure shows the stator voltage per phase, the stator current per phase, I_s, and the front-end converter current.

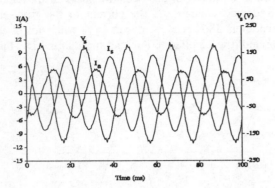

Figure 6. Steady state performance with inductive load.

5.3 Operation at sub super and synchronous speed.

Figure 7 shows the rotor current for subsynchronous operation (s≈0.1). Figure 8 shows the rotor current for supersynchronous operation (s≈-0.4). Both figures show that the rotor current distortion is low.

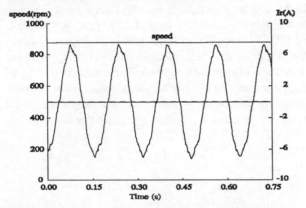

Figure 7. Rotor current at subsynchronous speed

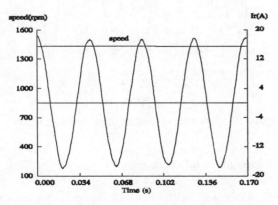

Figure 8. Rotor current at supersynchronous speed

Figure 9 shows the rotor current during a transition from subsynchronous to supersynchronous speed of the generator. The phase sequence of the rotor current changes when going through synchronous speed.

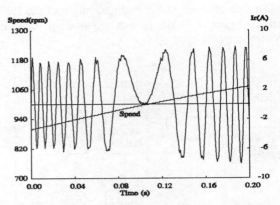

Figure 9. Transition through synchronous speed

CONCLUSIONS

The use of a DFIG to supply an isolated load has been presented. The system uses a Scherbius scheme with two voltage-fed current-regulated PWM converters allowing bidirectional power flow between the stator and the rotor. Operation at sub and supersynchronous speed with low distortion currents has been verified for both the machine and the front-end converter. The generator stator voltage has been regulated by controlling the stator flux magnitude. The control of the generator has been carried out using a d-q frame oriented along the stator flux vector positions. The control strategy of the front-end converter has been also carried out in a d-q frame aligned with the stator voltage position.

ACKNOWLEDGMENTS
R. S. Pena is grateful to the University of Magallanes and the Chilean Government for the financial support given to carry out this research.

REFERENCES.

1. Vicatos M S and Tegopoulos, 1989, 'Steady State Analysis of a Doubly-fed Induction Generator under Synchronous Operation'. IEEE Trans. on Energy Conversion, 4, 495-501.
2. Jeong S G and Park M H, 1987 'Steady State Analysis of a Stand Alone Wound Rotor Induction Generator Excited by a PWM Inverter'. IEEE IAS annual meet, 2. pp 790-797. USA 1987.
3. Bogalecka F, 1993 'Dynamics of the Power Control of a Double fed Induction generator connected to the soft power grid.' ISIE Budapest, 509-513.
4. Jones S.R., Jones R., 1993, IEE Colloquium on developments in real time control for induction motor drives", DIGEST No 1993/024, 5/1-5/9.
5. Leonhard W, 'Control of Electrical Drives'. Spring Verlag 1985.
6. Asher G.M., Sumner M., 1990. IEE Proc-D, 137, 179-188.

ADAPTIVE FUZZY CONTROL OF COMBINED CYCLE PLANT START UP

C. A. Smith, K. J. Burnham, P. A. L. Ham†, K. J. Zachariah‡, D. J. G. James

Coventry University, UK.
† Independent Consultant, UK.
‡Parsons Power Generation Systems Limited, UK.

INTRODUCTION

Greater thermal efficiency and flexibility of operation of combined cycle plant (CCP) have made their use more wide spread in recent years. A CCP makes use of one or more gas turbines, associated special boilers known as heat recovery steam generators and a steam turbine. The steam, for the steam turbine, is formed through utilisation of the waste heat from gas turbine(s) in the heat recovery steam generator(s). Further details of CCP theory, operational applications and mathematical models may be found in (de Mello et al (1); Helhofer (2); Scott and Russell, (3)).

The start up and shut down of CCP require the scheduling of gas turbines and the steam turbine to achieve the rated output in the shortest period of time. The most important variable to affect the start up time is the physical constraint of thermal stress. This paper will suggest the use of adaptive fuzzy logic in an attempt to optimise the start up and shut down times whilst keeping within thermal stress limits.

THERMAL STRESS

To attain the design life of a given machine the cumulative stress effects in the most highly stressed parts of the machine, i.e. the high pressure (HP) and intermediate pressure (IP) rotors must carefully be controlled. The life is mainly determined by the cycles of thermal stress experienced during the start up and shut down of the plant. Thermal stresses occur due to temperature gradients through the rotor metal. Important variables in the thermal stress equations include steam temperature, steam pressure and/or rates of change of these, heat transfer coefficient between steam and rotor surface and the dimensions of the rotor. Thermal stress equations are developed for a particular plant based upon a finite element grid to represent the inner temperatures of the rotor (Smith et al [4]; Young [5]).

OPERATIONAL ASPECTS OF POWER PLANT START UP

Two distinct stages are followed in the start up of steam turbine plant. The first stage, often called the 'run to speed' involves accelerating the turbine to the rated speed output. The machine is then synchronised to the grid and a small load (block load) is applied rapidly. The 'load up' that follows involves gradual application of the load on to the machine at various predefined rates until the desired output is achieved. According to traditional practice a fixed schedule is used for this process, which results in a predetermined time to completion; the choice of schedule is based on initial metal temperature levels. Such a procedure does not however allow shorter start up times to be used should the actual stresses prove, in fact, to be lower than predicted.

A basic problem arises due to the difference in timescales with the start up regimes of gas and steam turbines; the former may take less than fifteen minutes to achieve full load from a cold start, whilst the latter may well require two hours or more for the same situation. Additionally, however, load changing procedures after normal start up may also produce thermal stress cycles which should be considered during the operation of the plant. There are mechanical methods available for the control of the steam temperature, Smith et al (6) however, their use is limited and they are not always available. It should be noted that if such methods are available then they will complement the adaptive fuzzy logic control described in this paper.

It is a particular purpose of the current work to identify supervisory strategies which can be generalised to any plant configuration by making use of fuzzy logic. Other schemes for accommodating thermal stress in start up regimes are given in (Anderson (7); Dawson (8), Matasumoto et al (9)).

SUPERVISORY LOAD RATE CONTROLLER

The scheme's objectives are to ensure that rated output, i.e. full load, is achieved in the shortest possible time whilst respecting maximum stress

Opportunities and Advances in International Power Generation, 18–20th March 1996,
Conference Publication No. 419, © IEE, 1996

limits. The supervisory controller meets these objectives by altering the load rate, between the prescribed rates, through a decision making process. The decision making process is based upon fuzzy logic and decision regions (see references (4); (6); Smith et al (10); Smith et al (11)). The research reported in the aforementioned papers has been extended to encompass adaptation within a fuzzy logic scheme.

The fuzzy logic allows decisions to be taken on whether the load rate needs to be increased or decreased during the load up phase so that start up aims are met. The inputs to the fuzzy logic decision making system are the monitored stress, relative to its maximum, and its rate of change. The output, called the 'condition' will lie on the interval [-1,1] with the values '-1' and '1' indicating 'bad' and 'good' thermal stress states respectively.

To aid the decision making process future predicted 'conditions' and past 'condition values' are also included. Model based prediction of steam temperatures for n future samples allows the calculation of future stresses and hence future 'condition' values. These future values are averaged to give a 'future condition' value. Averaging m past 'condition' values gives a 'past condition' value. The average of the 'past', 'present' and 'future' subvalues gives a 'total condition' value.

The 'total condition' value is then compared with three decision regions: i)[1, TH_1,] ii)[TH_1,TH_2], iii)[TH_2,-1] being load regions where the load rate may be increased, kept the same and decreased respectively. Each prescribed load rate has an individual set of values for TH_1 and TH_2 and these are stored in an array. The values taken by the thresholds have previously been heuristically chosen (see references (4), (6), (10), (11)). Since the heuristically chosen values will in general offer a suboptimal solution an adaptive scheme is proposed.

ADAPTIVE THRESHOLD VALUES

To ensure convergence towards an optimal system some form of adaptation is required to be introduced. One form of adaptation considered is that of adapting the rule base of the fuzzy logic. However, the use of 'future condition' and 'past condition' values leads to a large number of rule changes. The approach is further complicated by similar 'total condition' values which have different implications. For these reasons attention is concentrated on adapting the thresholds. It is the aim of the adaptive fuzzy system to converge to an optimum set of plant specific threshold values. Therefore during early plant life there may be some adjustment of these values. As the threshold values

converge to their final steady state values the variance of threshold alterations will progressively decrease.

An integral aspect of the adaptation is the use of post run-time simulation. Post run-time simulation is used to assess the performance of each load rate change. If a decision failed to meet some criteria then simulation both back and forward in time is utilised. The aim of simulating around the decision point is to find a more appropriate time when a given load rate change should have been made. When 'looking forward' in time the system is required to be simulated as if no load rate change has been made. As a result, condition values used for simulation are predicted 'total conditions' rather than monitored 'total conditions'.

The 'total condition' value for the load rate change found by post run-time simulation, together with the corresponding load are stored in a knowledge base. A point of note is that each threshold value for each predefined rate will have an individual knowledge base, i.e. for two threshold values TH_1 and TH_2 and four load rates this leads to eight individual knowledge bases. The data stored in each knowledge base is used to form a relationship between the load and the threshold value. An assumption is made that the threshold value for each stored load data point should be equal to the corresponding stored 'total condition' value. The relationship is realised by fitting the equation of a straight line to the data by least squares. For the situation described in this paper there will be eight equations, one for each threshold value associated with each load rate.

The least squares algorithm gives the best parameter fit for a given equation and a given set of data points. The system equations for a given threshold value are as follows

$$y = \Phi\Theta \qquad (1)$$

where y is the vector of 'total condition' values, Θ is the vector of parameters for the straight line equation, i.e. a gradient value and a constant, and Φ is known as the observation matrix. Each row of the observation matrix contains the load value, for the corresponding 'total condition' in y, and a constant term of unity. The parameter vector which characterises the best straight line fit is then calculated from

$$\Theta = (\Phi^T\Phi)^{-1}\Phi^T y \qquad (2)$$

This adaptation procedure is carried out for each of the threshold values in the four load rates. Figure 1 illustrates one of the eight knowledge bases. The asterisk points on the graph are the 'total condition' value plotted against load. The solid line is the best straight line fit, calculated by least squares, through the points and is interpreted as the load dependent threshold values to be taken.

The following heuristic rules are included so that the threshold values are based upon interpolation rather than extrapolation. a) A decision may only be made if the present load is greater than the smallest load in the knowledge base for the load rate being used and b) the threshold value is held constant once the load is greater than the largest load in the knowledge base.

SIMULATION STUDIES

Simulation studies have been undertaken in order to demonstrate the adaptive threshold scheme. The model used is that of a CCP conversion supplied by Parsons Power Generation Systems Ltd. A total of twenty start ups have been simulated to form the knowledge bases, from which the thresholds are then created. In each of the training cases the initial metal temperature is altered in the range 20-200 °C. The thresholds are adapted so that the system would aim 2% below the maximum stress limit (i.e. 0.98 compared to a normalised maximum stress limit of 1). To test the effectiveness of the adapted thresholds fifty randomly chosen start ups within the temperature range of the training set are simulated and the maximum stress and time taken to achieve rated output recorded. The start ups are chosen so that approximately 50% of them give rise to situations where, in the absence of the fuzzy logic system and no operator intervention, the maximum stress would have been violated.

Table 1 illustrates the statistical results from these start ups. The stress is normalised in comparison to the maximum stress limit and the time is normalised with respect to the time taken with no fuzzy logic system. The results show that the mean stress value is close to the 98% requested and the standard deviation is approximately 2%. It should be noted that none of the start ups violated the stress limit and the stresses outside of the standard deviation were below the mean value. Considering how the scheme benefits the total start up time, it can be seen that the mean is approximately 90% hence an average saving of 10% in time. It is interesting to note that this saving in time is achieved despite the fact that approximately half of the start ups included a 'hold' on the start up in order to avoid overstressing the machine.

Figure 2 illustrates one such simulated start up. It contains three subplots where subplot a) represents a normalised thermal stress curve, b) the load rates used numbered 1 to 4 and c) the condition value with upper and lower thresholds, the condition value (solid line) and the thresholds (dashed lines). It may be seen that to avoid an overstress situation the load rate is

reduced to rate 1. Once the threat of overstressing is reduced the load rate is increased through the predefined rates.

CONCLUSIONS

This paper has demonstrated the use of a fuzzy logic based decision making system with adaptive thresholds to ensure the integrity of the life of highly stressed parts in a steam turbine. Realistic simulation studies have been undertaken to adapt and to test the fuzzy logic system using simulated data representative of a range of cold starts. Statistical analysis of the simulation results illustrate that plant integrity is maintained and the average time of start up reduced.

Future work is to extend the adaptation process to cope with changes in the steam temperature caused by variations in the gas turbines(s) start up.

ACKNOWLEDGEMENTS

The authors would like to thank the directors of Parsons Power Generation Systems Ltd for their support and permission to publish this paper. Also the EPSRC for the funding of this project.

REFERENCES

1. De Mello F. P. et al, 1994 "Dynamic models for combined cycle plants in power system utilities", IEEE working group on prime mover and energy supply models for system dynamic performance studies, Wm94-021.

2. Hehlhofer R. H., 1990 "Combined-cycle gas and steam turbine power plants" , Prentice-Hall, UK.

3. Scott P.H. Russell J.P., 1994. "Modelling of combined cycle conversion using SIMULINK." IEE Control '94, 1, 368-373, Warwick, UK.

4. Smith C.A., Burnham K.J., Ham P.A.L., James D.J.G., 1994 "Accommodating thermal stress during the start up of a combined cycle plant - A fuzzy logic approach." Tenth International Conference on Systems Engineering (ICSE '94), 2, 1151-1158, Coventry University, UK.

5. Young W.C.(1989) "Roark's Formulas for Stress and Strain." (6th Edn), McGraw-Hill.

6. Smith C.A., Burnham K.J., Ham P.A.L., Manchester J. R., James D.J.G., 1995 "Adaptive control of combined cycle gas turbine start up", IMechE seminar "The two-Shifting of CCGT and

Fossil Fired Steam Plant", (to be presented 4 December), London.

7. Anderson K., 1991. "Advanced control techniques reduce power plant stress" , Power, (May) 53 -56.

8. Dawson R. T., 1991 "The optimisation of operational flexibility through thermal stress monitoring and control." IMechE seminar "Load cycling, Plant transients and off line operation", 7-15.

9. Matsumoto H., Eki Y., Nigawara S., Tokuhira M., Suzuki Y., 1995. "An operation support expert system based on on-line dynamics simulation and fuzzy reasoning in fossil fired plants" IEEE Transactions on Energy Conversion, 8, 674-680.

10. Smith C.A., Burnham K.J., Ham P.A.L., James D.J.G., 1995. "Fuzzy and model based supervisory controller for combined cycle power plant start up using thermal stress constraints," Third European Control Conference (ECC '95), 3, 2444 -2449, Rome, Italy,

11. Smith C.A., Burnham K.J., Ham P.A.L., James D.J.G., 1995 " Optimising the start up schemes of combined cycle power plant", 2nd IFAC Symposium on Control of Power Plants and Power Systems (SIPOWER '95), (to be presented 6-8 December), Cancun, Mexico.

Table 1

	mean	standard deviation
Stress	0.9732	0.0204
Time	0.8950	0.0729

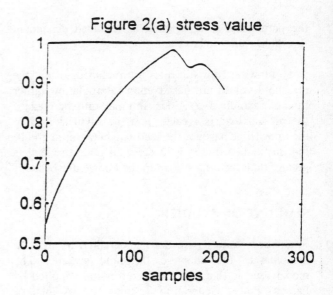

Figure 2(a) stress value

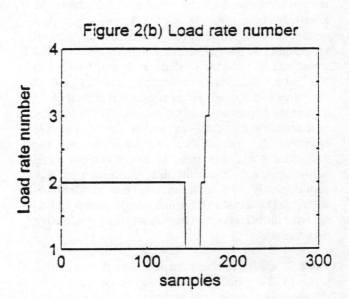

Figure 2(b) Load rate number

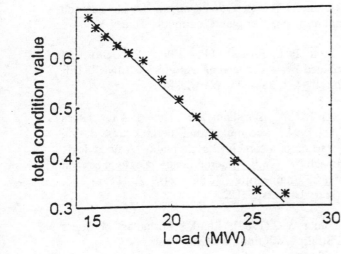

Figure 1

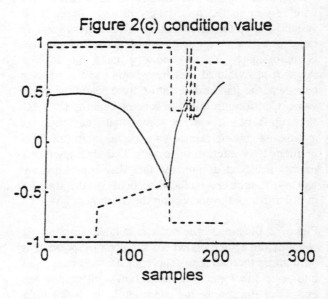

Figure 2(c) condition value

OPPORTUNITIES FOR ADVANCED SENSORS FOR CONDITION MONITORING OF GAS TURBINES

R M Cotgrove and M I Wood

ERA Technology Ltd., England

INTRODUCTION

The power generation business is currently a very competitive market for both the original equipment manufacturers (OEM) as well as the power utilities. A variety of economic and technical forces have produced large industrial gas turbines with ever increasing levels of efficiency, and operators are now being offered combined-cycle gas turbine plant with efficiencies of 57-60%. Simultaneously the operating companies are under continued pressure to minimise operating and maintenance costs. Since fuel costs represent the main part of the running costs, high efficiency units are attractive, providing they can meet the projected reliability and availability targets. However, many of the newer, larger, unit designs, despite proclamations of proven technology, involve a degree of technical advance and development over previous units particularly in regard of the use of higher firing temperatures and the design of the cooling systems for the first stage blades and nozzles. Part of the uncertainty as to the likely behaviour of the new designs arises from the OEM's current inability to test the large units at full load, except at a power station, with the consequential absence of extended running trials before entry into service. It is in fact the utility operators who generate the initial service experience on newer unit designs.

To minimise operator risks it is prudent to seriously consider the level of monitoring a unit requires, and whether this should go beyond that normally offered. It is within this context that the role of sensors and the associated opportunities lie. They provide a fuller picture of the operating conditions within the unit, allowing a move towards a condition monitoring based decision making process for both operational and maintenance issues.

This paper provides a review of advanced sensing techniques for monitoring the condition of gas turbines with particular emphasis on those parameters likely to be affected by the design changes found in the newer units.

MONITORS FOR GAS TURBINES

The Purpose of a Monitor

It is first worth considering briefly the purpose of a monitor. It receives information from sensors, and warns, instructs, or even takes action to change the system being monitored. The reason for implementing a monitor on a gas turbine, in common with any equipment, is to ensure that the performance or integrity of the unit does not fall outside acceptable limits. Possible objectives for monitoring include:

-To alert the operator of component failure to minimise the risk of death or injury, or of secondary damage to the unit and plant

-To allow the deteriorating component or system to be identified prior to catastrophic failure and hence to initiate appropriate action

-To allow the rate and nature of deterioration to be determined

-To allow the deterioration to be managed by modification to the operation of the unit to preclude catastrophic failure, extend life, and optimise the timing of outage maintenance.

-To allow the operator to confirm that the component or unit are performing as expected e.g. turbine rotor entry temperature

-To demonstrate compliance with legal requirements e.g. emissions control.

It is clear from these examples that the scope and nature of the sensors fitted to the unit, and the capability of the monitor to analyse, interpret, and take appropriate action will determine whether the monitor is a useful tool for predictive maintenance, or whether it acts solely as a fail-safe device.

Classification of Monitoring - Function & Timing

The classification of monitoring activity by function and timing helps to identify the requirements for any

Opportunities and Advances in International Power Generation, 18–20th March 1996,
Conference Publication No. 419, © IEE, 1996

monitor in relation to the system to be monitored. The fundamental nature of the system and the consequences of failure will determine the type and timing of the response from the monitor.

When critical components have failed it is imperative that a failure alert is given and that immediate action is taken . This is to ensure containment of any failed components, to preserve the integrity of what is essentially a high pressure combustion vessel, and to reduce the secondary damage caused either by mechanical damage or by fire. This type of monitor is required to be robust, and reliable. It is unlikely to have analysis capability, represents the minimum level of monitoring to allow 'safe' operation of the plant, and is probably of low cost. However, it may not present particularly good value for money when the consequences of a failure are considered in terms of lost production, damage to plant, and additional maintenance cost.

A monitor may be introduced for life management to attempt to preclude the possibility of failure. This may be to detect the onset of the stages of component failure, and to allow imperative action to be taken in the short to medium term to avoid unit failure whilst in operation. If the nature of the failure mechanism is understood it may be possible to modify unit operation to allow it to be run on to the next scheduled outage, or until an outage can be re-scheduled.

In the medium to long-term, monitoring for life management can be applied to optimise maintenance to minimise life-cycle cost. To complete the cost equation it is also necessary to monitor for operations management e.g.. by optimising the overall performance of plant comprised of multiple part loaded units. The monitors required for life and operations management require additional sophisticated sensors, frequently not provided as standard equipment, together with an increasing level of analysis capability, ranging from sophisticated data processing, to mathematical models which describe the integrity of components as a function of operating condition, or mimic the operation of the unit to allow performance optimisation to be performed.

The installation of specialist sensors may require modification to the unit. The capital cost of such monitors will be significantly higher than for simple failure alert system. However, the potential benefits, in terms of reduced life cycle, cost may outweigh this.

Characteristics of Monitors

Off-line monitoring is conducted when the unit is not running. Internal access is required to view or sample the monitored item, and the analysis of data, and presentation of the results is periodic. The analysis may take place on-site, or at a remote laboratory, and there may be a delay before results are available for interpretation and action. In contrast, on-line monitoring is conducted whilst the unit is running, and may be continuous or periodic The analysis and results update may be on-line in real-time, on-line and periodic, or even conducted periodically at some remote terminal or site. Monitors may be open-loop and advisory in nature, or may initiate a mandatory control system response if closed-loop. The latter require on-line real-time data analysis which is the most complex and costly type to implement.

Current on-line Techniques

Sensors and Transducers. If it is feasible to do so, improving the scope and quality of unit instrumentation is probably the most effective way to detect undesirable component operating conditions, such as temperature, pressure, or flow. Unfortunately, sensors capable of withstanding the harsh operating environment of the gas turbine are costly, and their reliability, stability, and service life may be inadequate for production use.

Gas turbines are normally supplied with the minimum instrumentation required to achieve satisfactory control of the unit. When additional information is required, algorithms which model and predict operating conditions at impractical sensor locations may be used. However, where the error in prediction is high, and monitoring the actual conditions provides a reliable indication of engine health, then the installation of a suitable sensor in the region of interest should be considered. Alternatively, an existing sensor may be upgraded to improve monitoring performance.

A good example of a sensor application is the exhaust gas temperature monitor. This is implemented as a radial thermocouple array located in the exhaust duct. It can detect bulk turbine over-temperature, and unacceptable temperature distributions, caused by problems with the upstream combustion or expansion process.

Vibration Monitoring. Vibration monitoring is performed to detect mechanical instability affecting stationary and rotating components i.e. shafts, vanes, blades, combustor, casings, ducts, and bearings.

A transducer senses a components motion in terms of its relative displacement, or its velocity, or its acceleration. The signal is fed to a system providing either an alarm or trip, with the option of additional data logging to enable basic vibration trending, through to sophisticated vibration spectra analysis. It is important that the requirement for steady-state and/or transient monitoring is specified, as transient analysis is more demanding. When analysis is required, it can be in the form of a permanent integrated installation, or as a portable diagnostic system. It is probable that displacement sensors would be permanently installed in the unit, since these detect the relative movement of, say, a shaft and bearing housing. However, external case mounting of accelerometers is possible, and this can be done without internal access to the unit.

The success of predictive maintenance employing vibration monitoring for gas turbine is highly dependent upon appropriate sensor location, and the achievable signal-to-noise ratio. Specialist advice should be sought otherwise monitoring will only provide a warning after a severe failure mechanism is active.

Performance Monitoring. Performance monitoring is used to determine the aerothermal performance and operating conditions, within the unit and its major components, relative to the installed datum. This may allow one to determine whether the performance of a unit has degraded, whether there an imbalance in unit operating conditions, or whether the unit is not being operated to its full potential.

Implementation requires an appropriate level of instrumentation, data acquisition and logging, and an empirical or analytical model of the unit covering the range of operating conditions, whether steady state and/or transient. The model is fed the operating parameters, and either calculates the predicted performance of the installed datum gas turbine, or carries out a look-up of actual recorded datum values. Additional component operating parameters can also be predicted.

It is possible to assess component performance degradation and recovery following preventive maintenance, such as compressor cleaning, and also to optimise unit operation to maximise profit. However, significant effort is required to calibrate the monitor, either by generating the empirical model, or by aligning actual unit performance with the analytical model. Such monitors rely upon the accuracy of sensor inputs, although they may allow predicted operating conditions to be calculated where visibility has been lost due to sensor failure.

Lubrication System Monitors. Oil chemical analysis allows the chemical composition of lubricating oils to be determined, this is frequently implemented in what is termed a Spectrometric Oil Analysis Programme (SOAP). Routine or exceptional periodic sampling of lubricants is carried out, analysis taking place either in-house or by sub-contract. The results are recorded, and compared with a set of acceptance criteria by constituent element, inorganic or organic compound type, and quantity.

This is an effective monitor of lubricant degradation and contamination, and the nature of the contaminants may indicate or confirm their source. A prerequisite is to establish the reference composition for the lubricant, to confirm that it meets the specification, and to calibrate the monitor. Sample handling procedures also need to ensure that the sample is not contaminated, nor deteriorates prior to analysis.

A monitoring technique complementary with SOAP is the detection and analysis of suspended solid material. Lubricating oils conveniently transport solid material from their source to settling tanks, and onto the surface of filters. It is therefore possible to detect and determine the source of wear of components if the composition of this material is determined. Oil Debris Analysis is a technique which concerns analysis of rate, size and nature of all debris.

Specifically, magnetic material can also be detected by appropriate location and installation of magnetic particle detectors within branches of the oil return system. A routine periodic sampling frequency is established, and the quantity and size distribution of the debris is measured and recorded. The results are compared with a set of acceptance criteria, and an action plan giving an escalating scale of additional monitoring or maintenance requirements is implemented until the problem is resolved i.e. ranging from an increased sampling frequency, to immediate unit shutdown. Metallurgical and compositional analysis of the material may identify its source, and hence the failing component.

Pyrometry. Monitoring key hot section temperatures can enable performance optimisation by improved unit control, and also optimise the operation and maintenance of the unit in relation to hot section component life. Unfortunately, even high temperature thermocouples are unable to operate for extended periods at temperatures approaching 1300°C found at turbine rotor entry. Radiation pyrometers do have the capability, and durability for this application, and they have been fitted to a number of types of industrial unit.

It is first necessary to install a 'window' in the casing of the unit which gives a line-of-sight to the surface of the component of interest. The radiation pyrometer then installed and aligned. By using high speed data acquisition and processing with suitable traversal it is possible to measure the surface temperature distribution across individual blades, in a rotating turbine, with a single pyrometer. Alternatively, a fixed point can be chosen on each blade i.e. nominally that which operates at the highest temperature. A radial array of pyrometers can also be installed to give the temperature distribution of a row of stationary components i.e. nozzle guide vanes.

Radiation pyrometers have seen many years of operation in military aero gas turbines, providing critical temperature information for engine control. They are shortly to enter service in commercial aero engines for airline service. They are not commonly found in the power industry, although they are used for development work, and are being employed as key sensors in a number of gas turbine durability surveillance programmes, as described by Ondryas (1).

It is important to note that a radiation pyrometer measures component surface temperatures, not gas temperatures, and for engine control and performance monitoring a model is still required to predict the gas temperature. In addition, the surface temperature reading is affected by the emissivity of the radiating surface, and the opacity of the viewing 'window'. Therefore coating type, coating degradation and soot formation all need to be considered when calibrating and operating the system, especially over a period of time. However, there are considerable opportunities to be gained once this type of sensor is installed, as shall be discussed later.

Continuous Emissions Monitoring (Pollutants). In many countries it is a mandatory requirement to demonstrate compliance with emissions legislation to the regulatory body, represented in the UK by HMIP (2), and an emissions monitor is therefore required. In circumstances where the performance of a single unit is degraded, and the emissions limit is exceeded, the monitor can assist in managing plant operation to ensure overall compliance with routine or exceptional short term emissions requirements. The commercial benefit of this type of monitor is dependent upon the financial implications of a failure to comply, and whether a financial incentive, such as a 'carbon tax', exists to stimulate further emissions reduction. However, there is potential for engine health monitoring, based upon the volume and composition of compounds found in the exhaust gas stream.

Continuous emissions monitoring provides greater than 90% accuracy in measuring pollutants when calibrated regularly. Unfortunately, calibration may need to be carried out daily, some types are unreliable and costly to maintain, and they are of high initial cost. Predictive emissions monitoring is an alternative technique under development, and dealt with later.

Techniques Under Development

Acoustic Emissions Monitoring. In a gas turbine an acoustic emissions monitor can be usefully applied to detect sliding contact between rotating and stationary components e.g. blade and abradable liner, or roller element and track. However, there are no known on-line monitors in production use, and its potential appears to lie in off-line monitoring of component interference to aid engine assembly, or as a diagnostic tool. Unfortunately, considerable effort may be required to characterise the response of the sensor with the nature, and location of the problem, and its use may be restricted to 'go' versus 'no-go' decisions.

Gas Path Electrostatic Particle Detection. Gas path electrostatic particles can be detected by means of sensing rings located in the inlet and exhaust duct of a gas turbine, as described by Cartwright et al (3). These particles are comprised of wear debris from components operating in the gas path, caused by tip rub, erosion, cracking, or foreign object damage (FOD). Two sensor rings enable the monitor to distinguish between particles caused by foreign objects, and those caused by component wear. The technique has been successful in detecting the onset of failure several hours before catastrophic failure occurred. This technique was forecast to enter service on military helicopters in 1996.

Rotor Blade Proximity Detection. Proximity sensors mounted around the circumference of the casing can detect passing rotor blades in both the compressor and turbine. This enables the casing to blade tip clearance and hence blade tip wear to be measured, or partial blade loss to be detected. Alternative sensor location enables axial clearance between blade and vane rows to be measured. Suitable sensor design also enables the vibration modes of the blade tip to be observed. These techniques have all been demonstrated successfully in development environments, and could assist in managing the engine running point to avoid compressor surge in worn units i.e. those with large blade tip and seal clearances.

Predictive Emissions Monitoring. Predictive emissions monitoring is a technique being developed to give an improvement in the cost, reliability and performance over that offered by continuous emissions monitoring equipment. It characterises emissions by means of a model of the unit, which is periodically calibrated using a suitable emissions analysis kit. It is necessary to develop a reliable empirical or analytical model of emissions from each type of unit, and to demonstrate its performance to gain acceptance of the technique by the legislature. EPRI studies indicate that the technique is feasible, and that it may be more reliable than continuous emissions monitoring, as outlined by Stambler (4).

Integrated Approach

Few of the previous techniques are implemented in isolation as many are complementary. A typical scheme of condition monitoring for a gas turbine would include:

vibration

performance i.e. standard day corrections, component efficiencies, bleeds, exhaust conditions

wear i.e. magnetic and/or debris

lubricant contamination i.e. SOAP

lubricant properties i.e. viscosity

lubricant temperature

lubricant consumption rate.

However there are moves towards the development of more integrated packages which make use of the greater computing power that is now available. An example of this would be the system installed by EPRI in its advanced gas turbine durability surveillance programme (1). This brings together aerothermal modelling and vibration monitoring into a single system, along with pyrometry for metal temperatures in the turbine hot section.

A knowledge-base can also be established which offers expert guidance for predictive maintenance, and this may be integrated into the monitoring system. The TIGER project described by Milne (5), is a good example of work currently underway in this area.

Future Opportunities

With reference to the monitoring techniques that have been described, the future opportunities in sensing technology for gas turbines are seen to lie in three areas.

Firstly, in the development of durable sensors which can withstand the harsh operating environment whilst delivering high levels of reliability, stability, and accuracy. As a preference improved temperature sensing would seem to provide the greatest benefits in terms of improved unit control to optimise unit operation, and component life management to reduce

maintenance cost. These two combine to reduce unit life-cycle cost.

Secondly, in a modest increase in the scope and quality of instrumentation provided as standard fit. This will be prompted by a better understanding of the benefits to be accrued from this policy, and hence by operators specifying a greater scope of instrumentation for their unit from the outset.

Thirdly, by making greater analytical use of the information available from sensors. This could include areas such as performance optimisation or assessing the condition of hot section components.

Motivation for advanced sensing

All three of the above opportunities stem from a requirement to more actively monitor and assess the operating conditions within a unit so that the operation and maintenance requirements can be most effectively managed so as to minimise risk and cost.

The application of sensor data for improved performance modelling and optimisation may be conceptually acceptable to plant personnel because of their familiarity with the engineering fundamentals behind the approach. However, it ought to be realised that the same type of data can be used to provide an improved basis for decisions on component maintenance intervals or component lives, especially for the turbine hot section. Typically 40%-60% of the maintenance costs derive from the repair or replacement of the hot section components. Whilst lives are frequently assessed on the basis of some form of equivalent operating load formulation, the better use of sensor data could lead to the development of inspection and repair intervals based on 'actual' operating conditions. These 'actual' conditions may still be calculated conditions based on aerothermal and heat transfer/finite element analysis or they may use inputs from more direct sensors such as pyrometry. However, they would be based on a more direct summation of the 'actual' conditions rather than the semi-quantitatively based equivalent hour type of formulation, hence leading to a more realistic assessment of the components operating history. Once this is combined with a methodology for assessing the condition of the component itself, the user has a sound basis upon which to make maintenance decisions.

DISCUSSIONS AND CONCLUSIONS

It may be seen from this brief review of the status of current sensors and the developments which are taking place that it is inevitable that there will be an increasing requirement for the information that they can provide.

It is likely that there will be the greater use of that information in ever more integrated analysis systems dealing with performance, component integrity/life assessment, and general turbomachinery trouble shooting such as bearing wear, vibration, etc. This will require both the more widespread application of sensors, but also an increasing reliability of these items under the more arduous conditions demanded for direct reading of operating parameters of interest, such as local metal temperature.

Acknowledgement

This paper is published with the permission of ERA Technology Ltd.

References

1. Ondryas I.S. et al, 1992, "Durability surveillance programme on the advanced GE Frame 7F", ASME reprint 92-GT-334, American Society of Mechanical Engineers NY

2. Her Majesty's inspectorate of pollution, 1994, "Chief inspectors guidance to inspectors; environmental protection act 1990", Process guidance note IPR 1/2 (revised 1994). Combustion processes. Gas Turbines., HMSO, London, UK

3. Cartwright R A., Fisher C., 1991, "Marine gas turbine condition monitoring by gas path electrostatic detection techniques", AMSE reprint 91-GT-376, American Society of Mechanical Engineers NY

4. Stambler I., 1995, "Utilities testing predictive emissions monitoring as an alternative to CEM", Gas Turbine World, 25-4, 20-25

5. Milne R., 1995, "Continuous expert diagnosis : is the future so far away?", Modern Power Systems (MPS), 15-10, 19-22

INTEGRATED GENERATOR PROTECTION, MONITORING AND CONTROL

C. R. Hodgson

GEC ALSTHOM T&D Protection & Control, United Kingdom

INTRODUCTION

The development of numeric protection has been prolific. A numeric device can provide a number of protection and non-protection functions which are integrated into a single unit. The protection of feeders was the first area to benefit from the emergence of this type of multi-functional approach, which has become widely accepted. Generator protection, which traditionally uses a number of discrete protective relays to provide each protective function, lends itself to the application of an integrated numeric device.

The application of numeric devices for generator protection is now beginning to gain widespread acceptance,. The benefits of using this type of protection are detailed in this paper.

PROTECTION FEATURES

A number of protection functions have been identified as being common to a wide range of generators (see figure 1). These protection functions are provided in the example package. They can each be enabled, or disabled, as required to suit the application. Discrete relays can be implemented for applications which require additional protection functions. Isolated digital inputs to the numeric relay are used to monitor, and log, the operation of any external, discrete protective relays, or plant contacts.

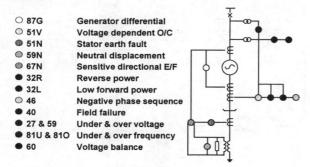

○ 87G	Generator differential
○ 51V	Voltage dependent O/C
● 51N	Stator earth fault
○ 59N	Neutral displacement
○ 67N	Sensitive directional E/F
● 32R	Reverse power
● 32L	Low forward power
○ 46	Negative phase sequence
● 40	Field failure
● 27 & 59	Under & over voltage
● 81U & 81O	Under & over frequency
● 60	Voltage balance

Figure 1 Generator protection functions

The operation of the protection functions and external inputs can be combined, through the use of internal scheme logic software, to control a number of output relays (see figure 2).

The provision of a number of protection functions and integral scheme logic within a single device results in a reduction in the time taken to design and construct the generator protection scheme. The physical size of the protection scheme is also greatly reduced.

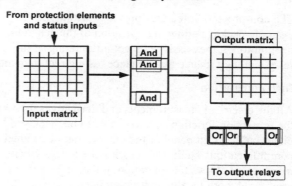

Figure 2 Internal scheme logic

Enhanced Protection Features

The protection which can be obtained using an integrated numeric relay has further benefits over a traditional scheme.

Frequency tracking. The sampling rate of the relay is synchronised to the running system fundamental frequency. This allows the protection to be maintained down to low frequencies, with adequate CT's and VT's applied. Thus the protection will be operative when variable frequency starting of the generator is employed.

Blocking inputs. The isolated inputs to the relay can be arranged to block the operation of certain protection functions. For example, a breaker auxiliary contact can be monitored to block the operation of the under voltage and under frequency protection whilst the generator is not connected to the power system. Blocking logic can also be used to co-ordinate the operation of back-up protection. The voltage dependant overcurrent protection and stator earth fault protection can be blocked by the operation of start contacts of down stream overcurrent and earth fault relays.

Dual settings. The relay has two independent groups of settings. A setting group change can be initiated remotely or locally, using dedicated inputs to the relay or by the operation of a protection function. This facility is useful for applications where the generator may be called on to operate in different modes, for example a pumped storage unit. Setting group selection is also a means whereby dead-machine protection can be introduced, when a machine is at standstill or on turning gear. With the second setting group selected protection against accidental energisation of the generator can be

Opportunities and Advances in International Power Generation, 18–20th March 1996,
Conference Publication No. 419, © IEE, 1996

provided by the voltage dependant overcurrent protection, this is set to operate instantaneously.

NON-PROTECTION FEATURES

Generator faults are not common, the protection will seldom, or never, operate during the lifetime of the plant. Protection must, however, be afforded. A digital device can provide additional functionality making the generator protection scheme more useful. The cost of protection can therefore be more justified, especially if the technology employed actually reduces the protection cost.

The non-protection features provided within the example package allow the operation of the generator to be monitored, they assist with commissioning of the relay and they reduce maintenance testing requirements

Instrumentation

All the electrical measurements and derivatives used by the protection functions are available to be displayed either on the front panel of the relay or remotely via the communications facility. Accessing the measurement information allows the operation of the generator to be monitored, both locally and remotely.

The measurement facility can be useful during the commissioning stage, to prove the connections between CT's and VT's and the protective relay. Periodic reference to the measurements can also be used to check that these connections remain intact.

Event, Fault and Disturbance Recording

Events can be the operation of any protection function, output relay or isolated input to the relay. A number of other events including change of setting group will also be logged by the relay.

Operation of any of the output relays can be arranged to create a fault record. A fault record consists of all the measurements at the moment that the output relay operates.

The operation of pre-defined output relays or isolated inputs is used to create a disturbance record. An oscillographical record of up to eight analogue channels, the status of the digital inputs and the relay outputs can be displayed using suitable viewing software.

Event, fault and disturbance records are invaluable if the generator develops a fault. Each of these records can be considered in determining how the fault occurred and developed. Appropriate repair work can then be quickly organised to minimise the outage period.

The fault records are not only available for generator faults however. The operation of external protection, for example line protection, can be arranged to operate an input to the device. This can be used to initiate the storage of a fault and disturbance record. The effect of an external fault, which would not cause operation of the generator protection, can be recorded and analysed.

Test Facilities

Test routines are provided within the relay. These allow the indicating LEDs to be cycled to ensure that they are operational. The scheme logic settings can be proved with a scheme verifier function. The output relays can be operated to test their functionality and the operation of any connected plant.

Self Monitoring

The relay conducts a diagnostic check on start up and continually checks its own functionality whilst in operation. The power-on diagnostic routine tests both the relay hardware and software. Successful completion of these tests enables watch-dog circuitry which continually monitors the relay hardware and software.

The self monitoring capability of the relay maximises availability by eliminating failures which may go unnoticed until the next time that the relay is tested.

If a failure occurs the watch-dog circuitry operates. This operates watch-dog contacts and the relay is taken out of service. The watch-dog contacts can be used to provide remote indication of relay failure, in order that remedial action can be undertaken. Where an outage of a generator may prove to be very costly, the provision of two protection schemes may be justified with automatic changeover between them initiated by the operation of the watch-dog contacts.

Remote Communications

Remote dialogue with the relay is possible over either a dedicated communications link or a multi-drop network. Both methods of connection allow access to the relay measurements, settings and fault, event and disturbance information

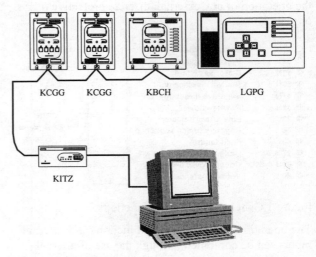

Figure 3 Communications access to various relay

The software used to access the relay is common to both methods of connection. When used on a multi-drop network it is possible to access a number of relays performing various functions.(see figure 3) The communication system is generic allowing similar communicating relays to be incorporated without continual modification to the access software.

The time taken to set a relay can be reduced by accessing the relay using the communication system. When setting the relay using the front panel, each parameter of the protection setting must be accessed individually. Accessing the relay via the communications system allows all the setting information to be displayed. This is demonstrated in figure 4. Also, where similar generators require similar protection, the settings from one relay can be extracted, using the communications system, and copied to the other relays.

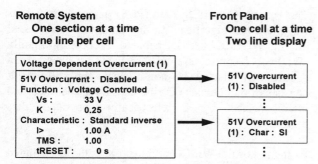

Figure 4 Comparison of access via communications and relay front panel

CONCLUSIONS

The primary duty of any protective relay is to afford comprehensive protection to the plant it is protecting. An integrated numeric generator protection relay supports a number of protection functions and internal scheme logic. The numeric design of the relay enables the protection to be further enhanced and a number of non-protection functions to be offered.

The non-protection features provide information on the status of the generator, during normal running and during faults, both locally and remotely via communications. Test routines, measurements and the communications facility assist in the commissioning of the device. Maximum availability of the protection is achieved through continual self monitoring.

THE IMPLEMENTATION OF THE CONTROL AND PROTECTION SYSTEMS FOR A NUCLEAR POWER STATION.

A Johnson BSc, MIEE, C.Eng
S Orme BSc, MIEE, C.Eng

Nuclear Electric Ltd, UK

The Sizewell B Control and Protection systems represent the most advanced implementations of such systems in an operating nuclear power station. The unique requirements of a nuclear power station have resulted in an approach to the control and protection systems which is both novel but applicable to any application where high integrity is a basic system requirement. The following discusses the implementation of the control and data processing system and reactor protection systems at Sizewell B.

CONTROL AND DATA PROCESSING SYSTEMS

At the centre of the Sizewell B control indication and alarm systems lie the Integrated Systems for Centralised Operation (ISCO). Early in the design of the power station it was determined to combine the control rooms, automatic control systems, plant control logic and data processing system into an integrated package known as the Integrated System for Centralised Operation (ISCO). Within ISCO were two main areas of supply; the control rooms supplied by CEGELEC-Pic and a data processing and control system supplied by Westinghouse Electric Corporation. This latter item became known as the Westinghouse ISCO (or WISCO) the engineering of which is now discussed.

The WISCO is a distributed system and consists of three major segments (see figure 1) each employing different control and instrumentation product lines; the High Integrity Control System (HICS), the Process Control System (PCS) and the Distributed Computer System (DCS).

Both HICS and PCS provide a multiplexed interface from control desk switches and indications to plant level actuators and sensors. The allocation of specific functions to each system is based on the nuclear safety categorisation of each control or indication.

The HICS provides a platform for the implementation of controls and indications which have the highest significance to nuclear safety. In addition it includes the Station Automatic Control Systems which, although less significant to nuclear safety, benefit considerably from the high integrity implementation. The HICS consists of 4 trains of computer subsystems which are located in electrically separated fire zones of the plant - a total of some 100 cubicles. Data is communicated between the various cubicles by means of redundant fibre optic data highways which are designed for high performance and

fault tolerance. The safety related manual and automatic control functions provided within the HICS are distributed between the 4 trains so as to provide the necessary degree of redundancy required to ensure safe operation in the presence of anticipated hazards such as fire.

The Process Control System (PCS) performs functions which are less significant to nuclear safety. It is based on the Westinghouse Distributed Processing Family (WDPF) of control system which has in excess of a decade of history in non-nuclear applications. It consists of 64 cubicles distributed around the plant and interconnected by 2 electrically separated data highways.

At the top of the WISCO architecture is the Distributed Computer System (DCS). This consists of 43 work stations networked together to access real-time plant data coming from the PCS, HICS or external datalink connections from other systems such as the Primary Protection System. This information is then presented to the operators in easily understandable form using over 450 plant mimic displays on graphic work stations. The DCS has no direct plant connection and does not perform any control functions but provides comprehensive alarm handling and logging facilities.

Design integration.

In total the WISCO has to handle over 2200 control modules in the main and auxiliary control rooms - representing over 8000 input and output signals. In addition the plant interface consists of approximately 20,000 signals resulting in a total of 28,000 I/O points which must be scanned. To accomplish this task over 200 cubicles (each typically consisting of I/O scanning logic and/or computing subsystems) are distributed throughout the power station. By relying on a distributed computing architecture the total function of the system is divided into sub-functions which can be individually optimised to achieve the best overall performance and reliability. The interconnection of these computers by a network of data highways allows them to share plant data without adding to the complexity of the cabling of the plant.

By their nature, the computer subsystems are defined by both their hardware and their software. In order to meet the demanding construction and commissioning programme a design approach was adopted for Sizewell

Opportunities and Advances in International Power Generation, 18–20th March 1996,
Conference Publication No. 419, © IEE, 1996

'B' which sought to separate the functional design of the system from the hardware design so as to permit procurement and installation of the hardware based on signal counts of the input sensors and control outputs together with a general estimate of the amount of processing power needed. The task of engineering the functionality of such a large system to the quality demanded of a nuclear installation was formidable and Nuclear Electric (in conjunction with CEGELEC-Pic) devised an object oriented approach to organising the applications data in order to manage the functional engineering effort. This methodology involved the conversion of baseline engineering documentation, produced by fluid system designers, into a consistent format from which control system engineers could configure the ISCO equipment. The majority of the functions which the ISCO was to perform were decomposed into a library of standard "models". The models defined generic processing requirements together with the man-machine and plant interfaces. Individual applications were subsequently defined by specific configuration and interconnection of models - then referred to as objects.

The object oriented approach was applied to all standard and repeated functions and resulted in approximately 16000 objects being defined. A few of the most complex control functions (eg the Station Automatic Control Systems) were still defined by traditional baseline engineering documents but the number of these was considerably less than the standard functions and thus more easily managed.

Hence, by adopting the object oriented approach to the design it was possible to represent most of the applications functionality within computer databases. In order to permit the skills of the software engineers to be focused on those areas which were unique rather than repetitive, extensive use was made of a variety of tools in the generation source code and configuration data for the computer subsystems. With the HICS, for example, tools were used which took as input the object oriented design data in the form of both databases and control logic graphics and converted this into source code or configuration data tables which were then linked to manually produced software modules.

The engineering design process consisted of a number of data translations. For example, input data to the process consisted of instrument schedules and schematics from fluid systems designers and contractors, this was translated into objects by the methodology mentioned above. Control system designers then translated the object oriented data into code for target computers whilst test and commissioning engineers translated the data into test procedures to demonstrate the accuracy of the design. At each translation stage the potential for error in the output existed. The incidence of such errors was significantly reduced both by the use of automated tools and by verifying the output of translations against the input.

The ability to use electronic databases in such verification activities significantly reduced the amount of human effort required whilst improving the integrity of the verification checks. As a final check it was general practice in the implementation of the WISCO to electronically compare the final source code for the target computers with the databases which originally went into the translation process.

The Sizewell B ISCO is the most advanced instrumentation and control system installed on a nuclear power plant to date. It represents the integration of a number of proven components into a network that provides a well engineered interface between the human operators of the plant and the plant equipment. A key feature of the distributed control systems is that they can be easily extended to embrace new functionality as the technology progresses or as plant experience develops. This flexibility allow Sizewell B to efficiently produce power well into the next century.

REACTOR PROTECTION SYSTEM

The Sizewell B Reactor Protection System (RPS) comprises two diverse systems; the Primary Protection System (PPS) and the Secondary Protection System (SPS). Diversity is a necessary part of the overall RPS design which results from the requirements of the station Probabilistic Safety Assessment (PSA). This requirement for diversity lead to the design specification for the PPS to be based on computer technology and the SPS to be based on existing solid state magnetic switching logic (the LADDIC technology).

The Primary Protection System is unique in that it is the only computer based UK nuclear reactor protection system used to protect against all design basis faults and its engineering is discussed in more detail below. By contrast the SPS uses conventional analogue signal processing and switching logic for it's functionality and is based on the LADDIC design which has had considerable operational experience on the later MAGNOX and all of the AGR reactor designs.

The functional scope of the two diverse systems resulted from an initial assessment of the station transient and fault analyses. The PPS functional design was based on the need to provide protection to mitigate all design basis faults. The SPS functional design was based on the protection necessary to mitigate faults with a frequency greater than once in a thousand years. Based on this broad functional definition the Engineered Safety Features (ESFs - automatically applied plant actuations) required to mitigate the faults within the design scope of each system were identified as were the input parameters necessary to initiate these ESFs. In addition, both systems are required to trip the reactor for their identified set of faults and hence each system has its own set of input parameters for this purpose. Inevitably, this means that in some cases common plant measurements are used by both the PPS and the SPS.

Where this occurs the actual measurement of the plant parameter is based on diverse techniques, thus preserving the diversity between the two systems.

Primary Protection System

The PPS is a computer based system providing both the reactor trip function and the initiation of the ESFs necessary to mitigate faults. The system input parameters are generally based on traditional sensors, mainly being RTDs, pressure transmitters and differential pressure transmitters. Digital techniques are used from the input processing through to the final voting logic for the reactor trip or ESF initiation functions. The use of digital techniques has significant advantages over analogue systems such as the ability to provide comprehensive diagnostic self testing and auto-calibration.

The PPS consists of four separate sets of equipment, known as trains, each is located in a different "Separation Group" to ensure segregation from the other trains. A particular parameter consists of four segregated sensors each providing an input into its own train. Each train provides a complete set of equipment for each input parameter and performs a comparison of each input against a setpoint in order to determine if the train is voting for a reactor trip or the initiation of the necessary ESFs. Optical datalinks from the other trains provide data to allow a two out of four vote to be performed in order to initiate the trip or ESF initiation. Optical datalinks were chosen for reasons of inherent immunity from electrical corruption coupled with high capacity. The basic functional logic is shown in figure 2.

All of the input parameters within a train are processed within the Integrated Protection Cabinet (IPC). This cabinet also performs the reactor trip voting and transmits parameter information to the WISCO (see earlier). Each IPC transmits the results of its setpoint comparison for the reactor trip function to the IPCs in the other trains and the results of the ESF setpoint comparison to the ESF Actuation Cabinets (ESFACs) in both its own train and to the ESFAC in the other trains. The ESFACs perform the voting required for the ESF initiation. In addition they receive the digital inputs from the Main Control Room for the initiation of the ESFs by the operators.

A further advantage of the use of digital systems is the ability to provide an automatic test system (the Autotester). The Autotester is used to carry out periodic functional testing of the system in order to meet its reliability requirements. Typically, the Autotester will be enabled on each train of the PPS in turn and will inject test signals into each parameter input - verifying that reactor trip or ESF actuation demands are generated at the correct signal levels.

The PPS has been implemented using individual microprocessor based sub-systems to perform functions such as performing a reactor trip or communications to the WISCO. Each sub-system uses a set of standard processor boards, the use of which allows a structured approach to the overall design of the software. In the case of a protection system the issue of the quality of the software is of key importance and was therefore was a prime consideration early in the design cycle of the PPS.

The structure of the sub-systems software encompasses a set of "Common Function " modules which are configured for the particular application by the use of "Configuration Data". The advantage of this approach is that the application code is greatly reduced as it does not have to consider the basic system functions which are handled by the common function modules. The common function modules are used in various applications and therefore have previous operating history.

The PPS was subjected to testing at each level during its design cycle. The design phase testing consisted of module testing which included type testing of boards and factory acceptance testing of a single integrated train. Once installed at site system integration testing was performed followed by commissioning tests. Testing at site also included a one year period of operation prior to being used in its protection system role.

The implementation of part of a reactor protection system using microprocessor technology is a major step forward in the UK nuclear power industry. Although microprocessor based protection has been used for protection system roles previously (such as fuel route protection systems for the AGRs) the PPS is unique in that it provides reactor protection for all the design basis faults. The PPS design incorporates many desirable features which result directly from the use of digital technology. Although the licensing associated with a software based protection system has been a time consuming and difficult process the was achieved and the plant has been operating commercially for many months as a result. The resolution of the licensing concerns has been assisted by the excellent performance of the system both during the commissioning phase and also during the early stages of commercial operation. This performance is a tribute to the designers, the system construction and the station commissioning team.

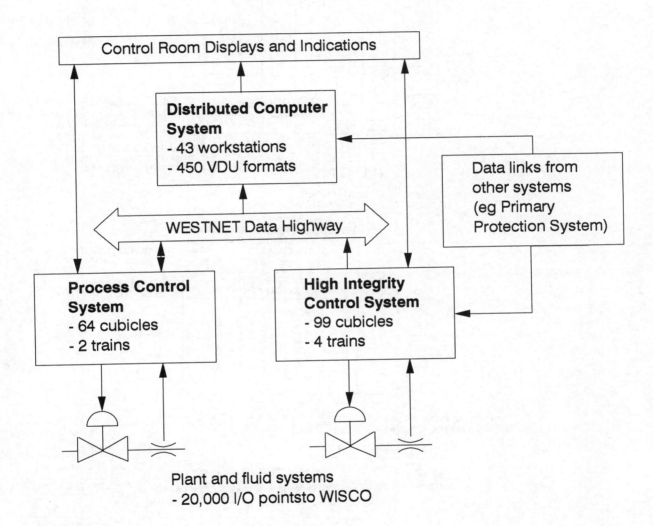

Figure 1: WISCO Principal Interfaces

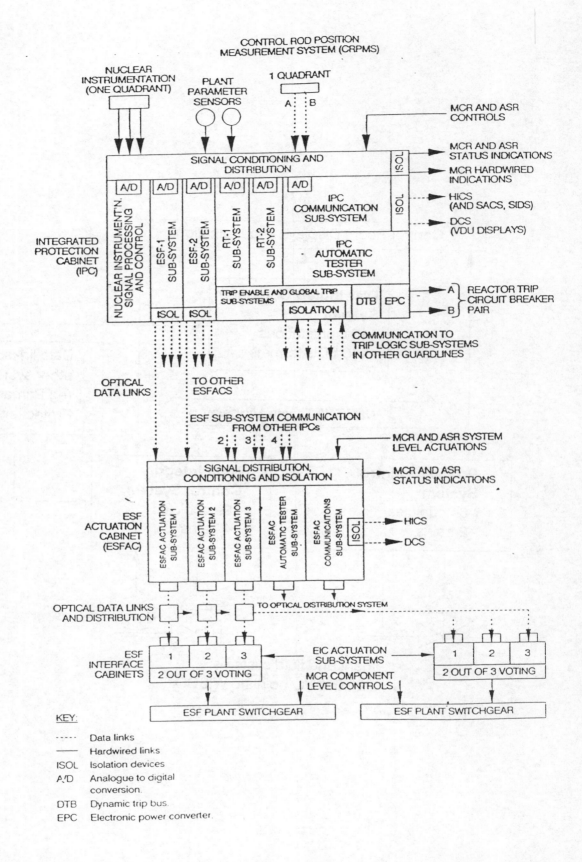

Figure 2: Subsystems in one PPS train

TRACING THE GENERATORS' OUTPUT

J. Bialek, D. B. Tam

University of Durham, UK

Abstract This paper introduces a simple topological method of tracing the flow of real and reactive power in transmission networks. The method may be used to allocate the charges for the reactive power, the transmission services and the transmission losses.

1. INTRODUCTION

The mesh structure of high-voltage transmission networks provides a large number of possible routes by which electrical power can flow from the sources (generators) to the sinks (grid supply points, here referred to as the loads). When the Electrical Supply Industry in England and Wales was privatised, the Electricity Pool Rules (1) stated that "with an integrated system it is not possible to trace electricity from a particular generator to a particular supplier" and this allowed to create a market place for electricity trading with profound consequences for the industry and the whole economy.

The Pool provides a market place for the real power trading but it does not solve the problem of charge allocation for a number of services necessary for a proper operation of the system, like use-of-the-system charge, losses, provision of reactive power, etc. Economical theory stipulates that this charging should be based on the marginal cost, Schweppe et al (2). It is widely known, however, that the marginal pricing is highly volatile, provides perverse economic signals to the transmission company and fails to recover the incurred costs, Dunnet et al (3). An alternative, or a supplement, to the marginal charging is a charge based on the actual system use, Shirmohammadi et al (4). However, due to a perceived impossibility of tracing how electricity flows in a network, it is often difficult to apportion equitably a particular charge to a particular network user. For example the transmission loss is covered at the moment by a uniform, pro-rata charge although certainly some users attract a higher loss than the others.

This paper endeavours to prove that, contrary to the common belief, it is possible to trace the flow of electricity in a meshed network. Consequently it is possible to create a table, resembling a road distance table, which shows what amount of real and reactive power is supplied from a particular source to a particular load. It is also possible to assess how much power provided by a particular generator flows in a particular line hence allowing to determine the network usage due to this generator. The proposed electricity tracing method may be used as a basis for improving the Pool system of settlements, especially that related to reactive power and transmission service charges, so that the real world is reflected in a better way.

2. ASSUMPTIONS.

The method proposed in this paper is topological in nature, in the sense that it deals with a general transportation problem of how the flows are distributed in a meshed, grid-like, network. The network is assumed to be described by a set of n nodes, m directed links (transmission lines or transformers), $2m$ flows (at both ends of each link) and a number of sources (generators) and sinks (loads) connected to the nodes. Practically the only requirement for the input data is that Kirchhohff's Current Law must be satisfied for all the nodes in the network. In this respect the method is equally applicable to real power flows, reactive power flows, or dc currents. Power flows satisfying Kirchhohff's Current Law may be obtained from a load-flow program or a state estimation program. The former case corresponds to analysing predicted power flows (short, medium or long term planning) while the latter corresponds to an after-event analysis of actual loading conditions.

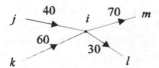

Figure 1 Proportional sharing principle.

The main principle used to trace the flow of electricity will be that of *proportional sharing* illustrated in Fig. 1 where four lines are connected to node i, two with inflows and two with outflows. The total flow through the node is $40 + 60 = 100$ units of which 40% is supplied by line j-i and 60% by line k-i. As electricity is indistinguishable and each of the outflows down the line from node i depends only on the voltage gradient and the line impedance, it may be assumed that each unit leaving the node contains the same proportion of the inflows as the total nodal flow. Hence the 70 units

Opportunities and Advances in International Power Generation, 18–20th March 1996,
Conference Publication No. 419, © IEE, 1996

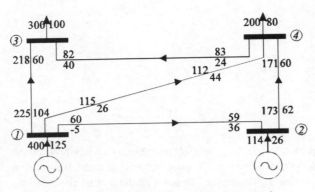

Figure 2 Ac power flow in the 4-node network.

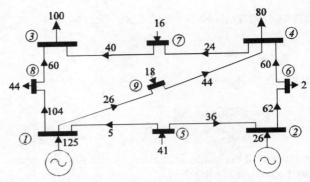

Figure 3 Reactive power flow corresponding to Fig. 2 with added fictitious line nodes.

outflowing in line i-m consists of $70\frac{40}{100} = 28$ supplied by line j-i and $70\frac{60}{100} = 42$ supplied by line k-i. Similarly the 30 units outflowing in line i-l consists of $30\frac{40}{100} = 12$ supplied by line j-i and $30\frac{60}{100} = 18$ supplied by line k-i.

The proportional sharing principle basically amounts to assuming that the network node is a perfect "mixer" of incoming flows so that it is impossible to tell which particular incoming electron goes into which particular outgoing line. This seems to agree with the common sense and with the generally accepted view that electricity is indistinguishable. The proportional sharing principle can obviously be neither proved nor disproved and hence it is of a little technical use, in the traditional, narrow sense of this word. It does, however, provide a powerful tool for setting charging mechanisms in the privatised and deregulated electrical supply industry.

3. TRACING ELECTRICITY.

Fig. 2 shows a simple 4-node network with active and reactive power flows. A number on top or to the left of the line indicates a real power flow, while a number below or to the right of the line indicates a reactive power flow. A similar convention has been used for the generators and the loads. The total transmission loss in the network is equal to the sum of all the line losses and equals $(225-218) + (83-82) + (173-171) + (60-59) + (115-112) = 14$ MW. The algorithm presented in this paper is referred to as *downstream-looking* as it looks at the flows outflowing from the network nodes. It is possible to devise a dual, *upstream-looking* algorithm analysing the nodal inflows, Bialek (5).

3.1. Application to the reactive power flow.

The values of the flows, both active and reactive, are different at the beginning and the end of each line. In order to trace where the output of each generator goes, it is necessary to introduce some fictitious "line nodes" responsible for the line loss. In the case of the real

power flow the line nodes act always as sinks whilst in the case of the reactive power flow the line nodes may act either as sinks or sources. This concept is illustrated in Fig. 3 which shows the modified reactive power flow from Fig. 2 and where nodes numbered from 5 to 9 are the fictitious "line nodes". Nodes 5, 7, and 9 act as the reactive power sources while nodes 6 and 8 act as the reactive power sinks.

The nodal flow through node i, Q_i, is equal to the sum of outflows

$$Q_i = \sum_{j \in \alpha_i^{(d,q)}} |Q_{i-j}| + Q_{Li} \quad \text{for } i = 1,2,...,n+m \quad (1)$$

where Q_{i-j} is the line flow from node i in line i-j, $\alpha_i^{(d,q)}$ is the set of nodes supplied directly from node i (i.e. reactive power must flow directly from node i to those nodes in relevant lines), and Q_{Li} the load at node i.

The line flow $|Q_{i-j}|$ can be related to Q_i by substituting $|Q_{i-j}| = |Q_{j-i}| = c_{ji}^{(q)}Q_j$ where $c_{ji}^{(q)} = |Q_{j-i}|/Q_j$ to give

$$Q_i = \sum_{j \in \alpha_i^{(d,q)}} c_{ji}^{(q)}Q_j + Q_{Li} \quad (2)$$

which can be re-written as

$$Q_i - \sum_{j \in \alpha_i^{(d,q)}} c_{ji}^{(q)}Q_j = Q_{Li} \quad \text{or} \quad \mathbf{A_{d,q}Q} = \mathbf{Q_L} \quad (3)$$

where $\mathbf{A_{d,q}}$ is the *downstream reactive distribution matrix* and $\mathbf{Q_L}$ is the vector of reactive nodal load demands. The (i,j) element of $\mathbf{A_{d,q}}$ is equal to

$$\left[\mathbf{A_{d,q}}\right]_{ij} = \begin{cases} 1 & \text{for } i = j \\ -c_{ji}^{(q)} & \text{for } j \in \alpha_i^{(d,q)} \\ 0 & \text{otherwise} \end{cases} \quad (4)$$

Application of equation (3) to the network shown in Fig. 3 gives

$$
\begin{bmatrix}
1 & 0 & 0 & 0 & 0 & 0 & 0 & -1 & -\frac{26}{44} \\
0 & 1 & 0 & 0 & 0 & -1 & 0 & 0 & 0 \\
0 & 0 & 1 & 0 & 0 & 0 & 0 & 0 & 0 \\
0 & 0 & 0 & 1 & 0 & 0 & -\frac{24}{40} & 0 & 0 \\
-\frac{5}{130} & -\frac{36}{62} & 0 & 0 & 1 & 0 & 0 & 0 & 0 \\
0 & 0 & 0 & -\frac{60}{104} & 0 & 1 & 0 & 0 & 0 \\
0 & 0 & -\frac{40}{100} & 0 & 0 & 0 & 1 & 0 & 0 \\
0 & 0 & -\frac{60}{100} & 0 & 0 & 0 & 0 & 1 & 0 \\
0 & 0 & 0 & -\frac{44}{104} & 0 & 0 & 0 & 0 & 1
\end{bmatrix}
\begin{bmatrix}
Q_1 \\ Q_2 \\ Q_3 \\ Q_4 \\ Q_5 \\ Q_6 \\ Q_7 \\ Q_8 \\ Q_9
\end{bmatrix}
=
\begin{bmatrix}
0 \\ 0 \\ 100 \\ 80 \\ 0 \\ 2 \\ 0 \\ 44 \\ 0
\end{bmatrix}
$$

$$(5)$$

Note that $\mathbf{A_{d,q}}$ is sparse and non-symmetric. If $\mathbf{A_{d,q}^{-1}}$ exists then $\mathbf{Q} = \mathbf{A_{d,q}^{-1} Q_L}$ and its i-th element is equal to

$$Q_i = \sum_{k=1}^{n+m} \left[\mathbf{A_{d,q}^{-1}} \right]_{ik} Q_{Lk} \qquad \text{for } i = 1,2,\ldots,n+m \qquad (6)$$

This equation shows how the nodal power, Q_i, is distributed between all the loads in the system. On the other hand the same Q_i is equal to the sum of all the inflows entering node i. A line inflow can be therefore calculated, using the proportional sharing principle, as

$$\left| Q_{i-j} \right| = \frac{\left| Q_{i-j} \right|}{Q_i} Q_i = c_{i-j}^{(q)} \sum_{k=1}^{n+m} \left[\mathbf{A_{d,q}^{-1}} \right]_{ik} Q_{Lk}$$

$$\text{for all } j \in \alpha_i^{(u,q)} \qquad (7)$$

where $c_{i-j}^{(q)} = \left| Q_{i-j} \right| / Q_i$ and $\alpha_i^{(u,q)}$ is the set of nodes supplying directly node i with reactive power. This equation shows a line flow as a sum of components supplying individual loads in the network. It is possible to devise a dual, *upstream looking* algorithm, Bialek (1996a), which would show a line flow as a sum of elements supplied from individual generators in the network.

The nodal generation is also an inflow and it can be calculated, using again the proportional sharing principle, as

$$Q_{Gi} = \frac{Q_{Gi}}{Q_i} Q_i = c_i^{(g,q)} \sum_{k=1}^{n+m} \left[\mathbf{A_{d,q}^{-1}} \right]_{ik} Q_{Lk} \qquad (8)$$

where $c_i^{(g,q)} = Q_{Gi} / Q_i$. This equation shows the i-th generation as a sum of components supplied to individual loads.

Let us now apply equations (7) and (8) to the network shown in Fig. 3. Inverting the matrix of equation (5) gives

$$
\mathbf{A_{d,q}^{(-1)}} =
\begin{bmatrix}
1 & 0 & 0.66 & 0.25 & 0 & 0 & 0.15 & 1 & 0.591 \\
0 & 1 & 0.139 & 0.577 & 0 & 1 & 0.346 & 0 & 0 \\
0 & 0 & 1 & 0 & 0 & 0 & 0 & 0 & 0 \\
0 & 0 & 0.24 & 1 & 0 & 0 & 0.6 & 0 & 0 \\
0.385 & 0.581 & 0.106 & 0.345 & 1 & 0.581 & 0.207 & 0.039 & 0.227 \\
0 & 0 & 0.139 & 0.577 & 0 & 1 & 0.346 & 0 & 0 \\
0 & 0 & 0.4 & 0 & 0 & 0 & 1 & 0 & 0 \\
0 & 0 & 0.6 & 0 & 0 & 0 & 0 & 1 & 0 \\
0 & 0 & 0.102 & 0.423 & 0 & 0 & 0 & 0 & 1
\end{bmatrix}
$$

Now the reactive power flow, say, $\left| Q_{9-1} \right| = 26$ MVAR supplies load *L3* with $(26/44) \times 0.102 \times 100 = 6$ MVAR and load *L4* with $(26/44) \times 0.423 \times 80 = 20$ MVAR. Generator, say, *G1* supplies load *L3* with $(125/130) \times 0.66 \times 100 = 63.5$ and load *L4* with $(125/130)0.25 \times 80 = 19.2$. The reactive power supplied from the other sources to the other sinks can be calculated in a similar way and Table 1 shows the resulting distribution of the generated reactive power:

TABLE 1 Distribution of reactive power.

Sources	Sinks				Total
	3	4	6	8	
1	63.5	19.2	0	42.3	125
2	5.8	19.4	0.8	0	26
5	10.5	27.6	1.2	1.7	41
7	16	0	0	0	16
9	4.2	13.8	0	0	18
Total	100	80	2	44	226

Table 1 resembles a road distance table and allows to trace how the reactive power flows from individual sources to individual sinks. This may have a number of useful applications as e.g. in allocation of charges for the reactive power. Obviously application of the algorithm to a large network would be cumbersome as a large matrix, of the order equal to $(n + m)$, needs to be inverted. Two observations could, however, ease the computational effort. Firstly, the matrix is highly sparse allowing the use of sparse inversion methods. Secondly, reactive power tends to flow rather locally so that it should be possible to split the network into several parts and investigate them separately.

3.2 Application to the real power flow.

The same electricity tracing method can be applied to the real power flow and the corresponding network diagram is shown in Fig. 4. All the fictitious "line nodes" act as real power sinks responsible for the line loss and Table 2 shows the resulting distribution of real power.

TABLE 2 Distribution of real power.

Sources	Sinks						
	3	4	5	6	7	8	9
1	267.36	120.36	1	0.68	0.6	7	3
2	32.65	79.63	0	1.32	0.4	0	0

This application of the electricity tracing method is especially important as it allows to apportion the total network transmission loss to individual generators. The output of *G1* used to cover the transmission losses is equal to 1 + 0.68 + 0.6 +7 +3 = 12.28 MW whilst the output of *G2* used to cover the losses is 1.32 + 0.4 = 1.72 MW. Hence the *net generation*, that is the actual generation minus the allocated loss, for generator *G1* is (400 - 12.28) = 387.72 MW and for generator *G2* is (114 - 1.72) = 112.28 MW.

As this method of calculating the transmission loss associated with a particular generator requires inverting a matrix of the rank ($m + n$), it is not computationally efficient and a method reported by Bialek (5, 6) suggests a simpler approach.

Another application of the proposed electricity tracing method regards allocating the network usage charge to the users. Note that equation (7), and its equivalent for the upstream-looking algorithm, allows to determine the share of a given generator or the load in a given line flow. Investigation is being carried out at the moment on how to use this, employing the well-known MW-MILE methodology developed by Shirmohammadi et al (7), to assess the total network usage due to a particular user.

4. CONCLUSIONS

In this paper a method of tracing the flow of electricity in meshed electrical networks has been introduced. The method results in a table, which resembles a road distance table, and allows to assess how much of the real and reactive power output from a particular generator goes to a particular load. It is also possible to calculate any line flow as a sum of components supplied from individual sources, or supplying individual sinks, in the network. When applied to the real power flow, the method allows to assess the transmission loss associated with each individual generator.

The proposed electricity tracing method may be used as a basis for improving the Pool system of settlements, especially that related to reactive power and transmission service charges, so that the real world is reflected in a better way.

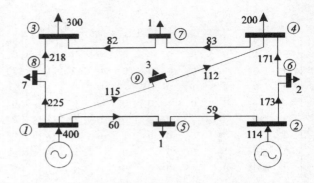

Figure 4 Real power flow with fictitious line nodes.

5. REFERENCES

1. The Electricity Pool of England and Wales, 1993, "An introduction to Pool Rules" prepared March 1991, updated December 1991 and April 1993, p.3.

2. Schweppe F.C., Caramanis M, and Tabors R.D., 1985, : "Evaluation of spot-price based rates" <u>IEEE Trans.</u>, <u>PAS-104</u>, 1644-1655.

3. Dunnet R.M., Plumtre P.H., and Calviou M.C, 1993, "Charging for use of transmission system by marginal cost methods" , <u>Proc. of Power System Computation Conference</u>, Avignon.

4. Shirmohammadi, D., Rajagopalan, C., Alward, E.R., Thomas, C.L., 1991, "Cost of Transmission Transactions: An Introduction" <u>IEEE Trans. Power Systems</u>, Vol. 6, No. 4, pp. 1546-1560.

5. Bialek J, 1996, "Tracing the flow of electricity" submitted to <u>IEE Proc.-Gener., Transm. and Distr</u>.

6. Bialek J, 1996, "Identification of source-sink connections in transmission networks" submitted to <u>IEE Fourth Int. Conf. on Power System Control and Management</u>, London, 16-18 April 1996.

7. Shirmohammadi, D., Gibrik, P.R., Law, E.T.K., Malinowski, J.H., O'Donnel, R. E., 1989, "Evaluation of Transmission Network Capacity Use for Wheeling Transactions", <u>IEEE Trans. Power Systems</u>, Vol. 4, No. 4, pp. 1405-1413.

THE DETECTION OF STATOR AND ROTOR WINDING SHORT CIRCUITS IN SYNCHRONOUS GENERATORS BY ANALYSING EXCITATION CURRENT HARMONICS

J Penman and H Jiang

University of Aberdeen, UK

INTRODUCTION

The synchronous generator is a principal elements of the power generation cycle. Its operational condition affects the electricity supply reliability and the system stability. With modern generators, protection and monitoring equipment are employed routinely to prevent damage. Rotor winding short circuits occur relatively frequently and if continuous, can lead to high temperatures and excessive vibration of the rotor resulting in damage to the generator if the fault is not cleared. To detect shorted turns in the rotor windings, several methods have been used for many years; for example, the airgap search coil technique, the monitoring of circulating stator current in double circuit machines, measurement the rotor shaft voltage, and the recurrent surge oscillograph (RSO) method, refs[1-3]

For stator winding shorts in double circuit machines circuits, measurement of the differential current between parallel connected half phases is used. Single winding stators present a more intractable problem, although it has been observed that there are some harmonic changes in line current under such conditions.

Here a novel method, which looks at the harmonic components present in the generator excitation current , is proposed to detect both stator and rotor winding shorts. Analytical justification of the method is given together with experiment results obtained using two synchronous machines of different ratings, both of which have single circuit stator windings and four pole, salient rotors..

In salient pole machines, when a short occurs in one pole, the mmf of this pole will change and the flux distribution in the airgap will no longer be symmetrical, This asymmetrical distribution induces additional harmonics in the stator winding and causes current harmonics to be generated in the rotor winding. These will circulate in the rotor winding and the exciter. Hence, by analysing exciter current harmonics, we can detect the rotor winding shorts. The same technique can also be used for generator stator winding shorts. When the stator has a shorted turns in one phase, a pulsating current results in the short circuit; and counter-rotating rotating magnetic fields are generated. One of these fields cuts the rotor winding and induce harmonics in the rotor which also couple to exciter, as before, and likewise offers a means of detecting such shorts.

SALIENT POLES MACHINES

Consider a four-pole salient-pole machine with a distributed winding stator, as shown in Figure 1. The concentrated rotor coils have 2N turns, carrying I amperes. The developed winding is given in Figure 2, together with the corresponding MMF wave.

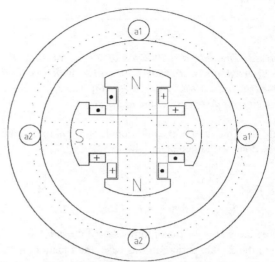

Figure 1 Four-pole Machine with a salient-pole rotor and a distributed stator winding.

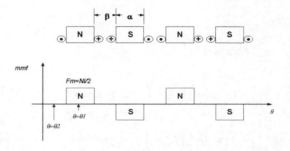

Figure 2 Developed rotor and corresponding mmf distributions.

A maximum mmf of value NI/2 occurs over the polar region. Fourier analysis, with zero position of θ taken as θ_2, gives:

$$mmf_\theta = \frac{4F}{\pi} \sum_{n\,odd} \left[\pm k_n \sin(np\theta) \right], \quad \text{with}$$

$$k_n = \frac{\sin\left(n\frac{\alpha}{2}\right)}{n^2 \frac{\alpha}{2}},$$

where F is the peak mmf of each of $2p$ poles..

When a rotor pole coil has a shorted turn, the total ampere-turns of that pole reduces. The mmf distribution on the pole with shorted turns can be shown in Figure 3a. Neglecting saturation, the effect of the shorted field coil can be considered as the normal field with an additional field due to a fictitious coil of the same number of turns as the shorted turns but with opposite current flow. See figure 3c.

Opportunities and Advances in International Power Generation, 18–20th March 1996, Conference Publication No. 419, © IEE, 1996

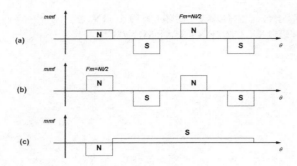

Figure 3. The mmf distributions of the rotor with one pole with shorted turns.(a) shows the rotor with one coil with shorted turns, (b).illustrates a healthy rotor and (c). is the assumed mmf distribution due to the short.

Analysis of the *mmf* of the additional rotor coil due to the short can be done with reference to figure 3c;

$$f(\theta_r) = \begin{cases} -f_m & for\ 0 \le \theta_r \le \dfrac{\pi}{4} \\ f_m' & for\ \dfrac{\pi}{4} < \theta_r \le 2\pi \end{cases} \tag{1}$$

where f_m denotes the maximum value of the *mmf*.
Noting the double period, in this case, the function, $f(\theta_r)$, can be expressed as follows:

$$f(\theta_r) = A_o + \sum_{n=1}^{\infty} \left[A_n \cos(n\theta_r) + B_n \sin(n\theta_r) \right] \tag{2}$$

where

$$A_o = \frac{1}{2\pi} \left[\int_0^{\frac{\pi}{4}} -f_m d\theta_r + \int_{\frac{\pi}{4}}^{2\pi} f_m' d\theta_r \right]$$

$$A_n = \frac{1}{\pi} \left[\int_0^{\frac{\pi}{4}} -f_m \cos(n\theta_r)d\theta_r + \int_{\frac{\pi}{4}}^{2\pi} f_m' \cos(n\theta_r)d\theta_r \right]$$

$$B_n = \frac{1}{\pi} \left[\int_0^{\frac{\pi}{4}} -f_m \sin(n\theta_r)d\theta_r + \int_{\frac{\pi}{4}}^{2\pi} f_m' \sin(n\theta_r)d\theta_r \right]$$

Equation 2 has the following space harmonics;

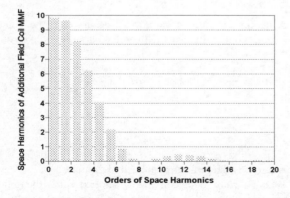

Figure 4 Space Harmonics Components of MMF produced by the fictitious additional field coil on the rotor.

Flux density distributions in the airgap. The airgap permeance distribution of a salient-pole machine, can be approximated as the sum of a constant component and a *(2p)^{th}* harmonic term rotating at synchronous speed, thus

$$\lambda_r = \left[\lambda_0 + \lambda_1 \cos(2p\theta_r) \right]$$

where λ_r denotes permeance of the air gap, λ_0 the constant component of the permeance and λ_1 the peak magnitude of the rotating component of the total permeance. The flux density is defined as
$B_r = mmf_r \bullet \lambda_r$, therefore, by using equation 2,

$$B_r = \left[A_o + \sum_n \left[A_n \cos(n\theta_r) + B_n \sin(n\theta_r) \right] \right] \bullet \left[\lambda_0 + \lambda_1 \cos(2p\theta_r) \right] \tag{3}$$

Which has the flux density harmonics shown in the Figure 5.

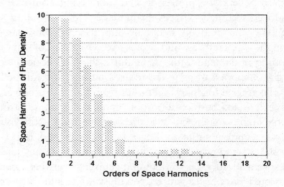

Figure 5. Space Harmonics Components of Flux Density in Air Gap.

We allow for the relative angular speed of the rotor with respect to the stator by $\theta_s = \theta_r + \omega_r t$, where θ_s is the angular position with respect to the stator and ω_r the angular velocity of the rotor. Equation 3 can then be rewritten as;

$$B_s = \left[A_o + \sum_n A_n \cos(n(\theta_s - \omega_r t)) + B_n \sin(n(\theta_s - \omega_r t)) \right] \bullet \left[\lambda_0 + \lambda_1 \cos(2p(\theta_s - \omega_r t)) \right] \tag{4}$$

Flux linkage to the stator winding. From the equation 4, we can see that the flux density distribution, in the stator frame, has both space and time harmonics, and all space harmonic components in the rotor frame exist in the stator frame.
The corresponding flux linkage with the stator winding can be expressed as:

$$\psi_s = \int_s B_s dA = \int_0^{2\pi} N_s \cdot B_s \cdot l \cdot r d\theta_s \tag{5}$$

where N_s is the stator winding distribution which is a function of angular position and given by:

$$N_s(\theta_s) = \sum_n k_n \cdot \sin(n\theta_s) \tag{6}$$

Thus,

$$\psi_s = \left[\int_0^{2\pi} \sum_n (k_n \cdot \sin(n\theta_s)) \cdot \left[A_o + \sum_n A_n \cos(n(\theta_s - \omega_r t)) + B_n \sin(n(\theta_s - \omega_r t)) \right] \right] \cdot \left[\lambda_0 + \lambda_1 \cos(2p(\theta_s - \omega_r t)) \right] d\theta_s \right] \tag{7}$$

From Faraday's law, the voltage induced in the stator winding is given by

$$E = -\frac{d\psi_s}{dt}$$

Induced emf in the rotor. The magnetic field of the rotor rotates in the airgap at synchronous speed; the stator windings cut this rotating field producing *emf*. If the windings is connected to a load forming a closed circuit, then this emf will produce current in the stator winding. The currents in the winding, flux, produce the armature reaction field which also rotates at synchronous speed inducing no emf in the rotor. However, when the rotor field coil has a shorted turn, symmetry is lost and the reaction field links with the shorted turn. From the foregoing analysis, we note that there are harmonic flux linkages with the stator due to the rotor shorted turn. These harmonics produce a corresponding rotating magnetic field in the stator winding rotating at the appropriate time harmonic frequency, and with a different relative speed with respect to synchronous speed. This speed difference will again induce emf in the rotor winding, consequently generating a series of alternating current components in the rotor. They are shown in the table 1.

Table 1 Harmonics Induced in the Rotor Winding(Hz)

Harmonic Components(Hz)								
In the stator	25	50	75	100	125	150	175	200	225
Syn. Speed	50	50	50	50	50	50	50	50	50
In the rotor	-25	0	25	50	75	100	125	150	175

EXPERIMENTAL RESULTS - ROTOR SHORTS

Figure 6 gives the stator current harmonics for the machine under normal conditions and with a shorted rotor. The changes are apparent, particularly for the 75Hz component.

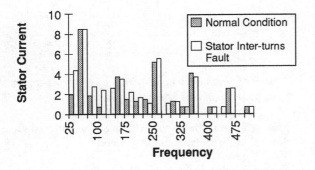

Figure 6 Stator current harmonics

Figure 7 shows the current harmonics in the rotor winding due to stator winding reaction.

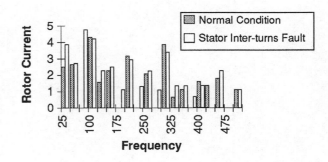

Figure 7 Rotor Winding Current Harmonics.

The machines under test here have a rotating excitation systems, and as reasoned previously it should be possible to detect the presence of rotor short circuits by inspection of the harmonic components in the exciter stator. These harmonics are shown in Figure 8.

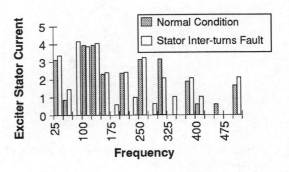

Figure 8 Exciter Stator current harmonics

From the above it can be seen that it may indeed be possible to detect rotor short circuits through the examination of exciter current harmonics.

Stator winding short circuits. The generator under test here has a stator winding with 36 slots total, 3 coils per group, two opposite groups in parallel form one section, two sections in series per phase, with each coil spanning 9 slots in a lap arrangement. Figure 9 shows the arrangement. one layer layout of the construction of the stator windings. When the winding has an inter-turn fault, the current in the shorted turns will change. Figure 10 shows the winding with shorted turns across three slots as used here. As argued previously, it can be seen that a stator winding short can also be considered as comprising the normal winding with an additional coil superimposed with the same number of turns as the shorted turns, but with opposite current flow. The current in this additional coil will produce additional flux in the air gap, but here, since the current is in a single phase, a pulsating magnetic field will result.

140

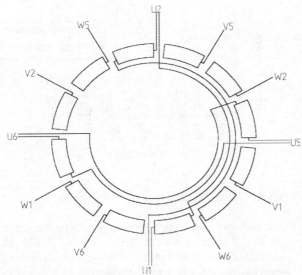

Figure 9 Stator winding layout ,one layer only shown.

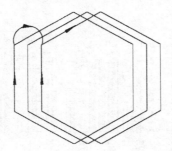

Figure 10 The diagram of shorted coils in one group of one section of a single phase.

Mmf generated by the additional coil. Figure 11 shows the mmf distribution of the superposed additional coil, and Figure 12 gives the frequency content of this distribution.

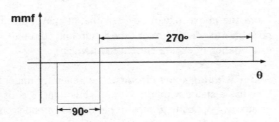

Figure 11 The mmf distribution of the additional coil.

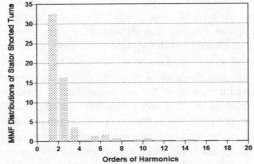

Figure 11 Frequency components of mmf distributions of the additional coil, stator reference.

The mmf distribution shown in Figure 11 can also be expressed by the following equation:

$$f(\theta_s) = \begin{cases} -f_m & for\ 0 \le \theta_s \le \dfrac{\pi}{2} \\ f_m' & for\ \dfrac{\pi}{2} < \theta_s \le 2\pi \end{cases} \quad (8)$$

This representation is similar to that of equation 1, except for the differences in range and the change to subscripts to represent position around the stator, rather than the rotor. The remaining analysis proceeds in exactly the same manner as before, and in this way it can be shown that the airgap flux density, relative to the stator winding, and in the presence of a stator interturn short can be represented thus;

$$B_s = \left[A_o + \sum_n \left[A_n \cos(n\theta_s) + B_n \sin(n\theta_s)\right]\right] \bullet \left[\lambda_0 + \lambda_1 \cos(2p\theta_s)\right] \quad (9)$$

Due to the relative angular speed between stator and rotor, $\theta_s = \theta_r + \omega_r t$, where θ_s is the angular position with respect to the stator and ω_r the angular velocity of the rotor, then equation 9 can be rewritten with respect to the rotor as;

$$B_r = \left[A_o + \sum_n \left[A_n \cos(n(\theta_r + \omega_r t)) + B_n \sin(n(\theta_r + \omega_r t))\right]\right] \bullet \left[\lambda_0 + \lambda_1 \cos(2p(\theta_r + \omega_r t))\right] \quad (10)$$

The corresponding frequency components are shown in Figure 12, below.

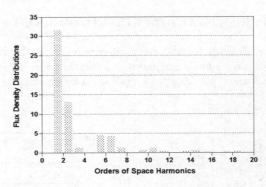

Figure 11 Frequency components of mmf distributions of the additional coil, rotor reference.

Flux linkage to the rotor winding. The corresponding flux linkage with the rotor is therefore:

$$\psi_r = \int_s B_r dA = \int_0^{2\pi} N_r \cdot B_r \cdot l \cdot r d\theta_r \quad (11)$$

where N_r is the rotor winding distribution, which is the function of rotor angular position and can be expressed as:

$$N_r(\theta_r) = \sum_k s_k \cdot \sin(k\theta_r) \quad (12)$$

Hence;

$$\psi_r = \left[\int_0^{2\pi} \sum_k \left(s_k \cdot \sin(k\theta_r) \right) \cdot \left[A_o + \sum_n A_n \cos(n(\theta_r + \omega_r t)) + B_n \sin(n(\theta_r + \omega_r t)) \right] \cdot \left[\lambda_o + \lambda_1 \cos(2P(\theta_r + \omega_r t)) \right] \theta_r \right]$$

(13)

Induced emf in the rotor. The use of equation 13 is very difficult if the appropriate details of rotor winding distributions are not known. Here, the machine under test had a rotor winding of four coils separately wound on four poles The coil pitch is around 45 degrees and the connections between each coil are as shown in Figure 12.

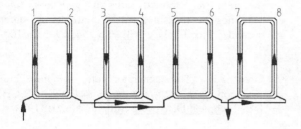

Figure 12 Rotor winding connections

To help understand how the rotor coils link with the fluxes defined by equation 13, it is best to re-draw the rotor coil positions and distribution, as shown in Figure 13. Here, each smaller circle represents one coil side and they are serially connected in the manner shown in the Figure 12. The rectangles indicate the poles of flux density distributions due to stator winding in the normal condition, and due to the fictitious coil representing the interturn fault.

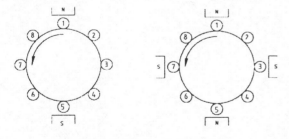

Figure 13 Simplified machine layout for stator interturn faults.

From the results of Figure 11, it is clear that the first and secondary order space harmonic components dominate the flux density distribution.

For the first order space harmonic, the equivalent machine is shown on the left in Figure 13. The sides of the rotor coils will cut the flux produced by the stator winding. For each revolution each conductor cuts flux once. When No1 conductor is at the position shown in Figure 13, the directions of induced emf in conductors No1 and No5 will be opposite, as it will be for Nos 2&6, 3&7, and 4&8. However, from Figure 12, we note that for these pairs of conductors the current flow should always be in the same direction. This indicates that the emf induced by the first order space

harmonic in these conductor pairs will cancel each other, and there will be no induced current due to this component.

For the second order space harmonic component, the equivalent machine is shown on the right hand side of Figure 13. Here, each conductor will cut flux twice for each rotor revolution. The appropriate conductor pairs in this case are Nos 1&3, 2&4, 3&5, 4&6, 5&7, 6&8 and 7&1 respectively. For the position as shown in Figure 13, the induced emf in conductors of 1 and 5 have the same direction, whilst 3 and 7 have emf induced in the opposite direction to that in 1 and 5. With this arrangement the emfs induced in each of these conductors will produce current in the rotor, and the direction of the current is that shown in Figure12. Because each conductor cut the flux twice for each rotor revolution and the eight conductors are connected in series, the emf induced in each conductor will change direction twice for each revolution, so that the total emf generated acoss all eight conductors will change direction 16 times with each revolution. Since the rotor angular speed is 25Hz for a 4-pole machine, the frequency of the current induced in the rotor winding by the stator winding fault will be 16×25Hz=400Hz.

EXPERIMENTAL RESULTS - STATOR SHORTS

As predicted above, Figure 14 shown that the 400Hz component of current in the rotor winding changes significantly when the stator winding has an inter-turn short circuit, verifying our prediction.

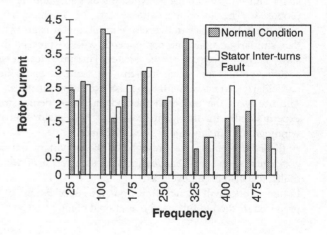

Figure 14 Rotor current harmonics due to stator interturn short circuit.

Since the rotor winding, particularly for a brushless machine, cannot be accessed easily whilst the machine is operating, it is very difficult to measure the rotor current without first modifying the machine. It would therefore be helpful to be able to detect variations in the rotor current harmonics by measuring the exciter stator current. As discussed previously when considering rotor faults, this is entirely possible since the same harmonic current components as appear in the rotor winding should also appear in exciter stator winding. Figure 15 confirms this and shows too the large change in the 400Hz component due to the presence of the fault in the stator winding of the generator.

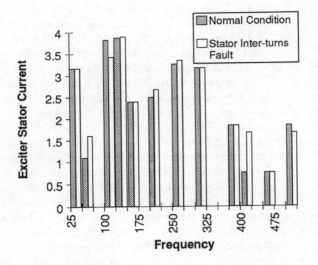

Figure 15 Exciter stator winding current harmonics due to an interturn short circuit in the stator winding of the generator.

CONCLUSIONS

A novel method of detecting short circuits in the both the stator and rotor windings of synchronous generators has been proposed. The technique allows for the identification of predictable harmonics in the rotor current, and it is also shown that similarly predictable components will be present in the stator current spectrum of the exciter. This provides an easier access point for measurement of these components, particularly generators with brushless excitation schemes.

The utility of the proposed method has been demonstrated, experimentally for the identification of both stator and rotor winding faults in laboratory sized machines.

A theory explaining the presence of the important harmonic components, and the changes induced as a result of winding faults, has also been developed.

Future work will concentrate on the application of the proposed method to full sized operational plant.

ACKNOWLEDGEMENTS

The authors would like to express their appreciation to the EPSRC and the University of Aberdeen Development Trust for the support of this work.

REFERENCES

1. Conolly, HM, Jackson RJ, Lodge I and Roberts IA; "Detection of shorted turns in generator rotor windings using airgap search coils" Proc EMDA, IEE Pub 254, 1985, pp11-15

2. Muhlhaus J, Ward DM, and Lodge I; "The detection of shorted turns in generator rotor windings by measurement of circulating stator currents" Proc EMDA, IEE Pub 254, 1985, pp100-103

3. Grant AE; "Turbogenerator rotor winding fault detection by a recurrent surge method" CEGB, UK, Technical Disclosure Bulletin, No201, 1973

EXPERIENCE WITH VRLA CELLS

K S Pepper, M C Barkwith and D M Ward

Nuclear Electric plc., UK

Until recent years, the batteries used in our Power Stations for No-Break Essential supplies have predominately been of the flooded lead-acid Planté design. However, the last decade has seen a new contender enter the marketplace, the Valve Regulated Lead-Acid, or VRLA, cell, also known as the sealed lead-acid or gas recombination cell. We alone have in excess of 7500 of them installed at Sizewell B. They are used for tasks as diverse as fire detection equipment and safety UPS systems, ranging in size from 24V 6.5Ah to 110V 1200Ah or 350V 800Ah. This paper deals with the design aspects of the VRLA cell and also the hidden dangers.

Looking first at the design of a VRLA cell, see Figure 1. The overall design varies little between different manufacturers. The positive and negative plates within the cells are formed from a grid of lead alloy pasted with a mixture of lead oxide, sulphuric acid, together with other additives. The plates are kept apart by micro fibre glass mats or gel, which together with the plates absorb the acid in their pores. As a result, there is no free acid.

As with all lead acid cells, if a current is passed through a fully charged cell, the water content is electrolysed liberating oxygen at the positive plate and hydrogen at the negative plate. A VRLA cell differs from a normal Planté

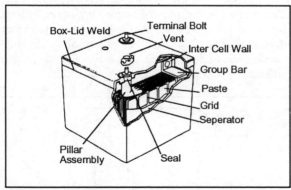

Figure 1 : Design of a VRLA cell

cell in that the oxygen formed at the positive plate diffuses to the negative plate where it reacts to form a lead oxide. This then reacts with the sulphuric acid electrolyte to form lead sulphate and water. The lead sulphate is then electrochemically converted back to lead by the passage of a charging current. As a result the oxygen evolution at the positive plate is balanced by the reduction of lead sulphate to lead as opposed to hydrogen evolution.

To prevent discharge of the negative plates by atmospheric oxygen, the cells are operated at between ½ and 5 psi above atmospheric pressure. A valve is provided in the top of the cell, which in addition to preventing oxygen ingress, also relieves any excessive pressure inside the cell. These valves typically consist of a valve body moulded as part of the lid, a neoprene nitrile rubber valve cap, a porous flame-arresting disc, and a clip-on valve retainer.

Since there is very little water escape from a VRLA cell by electrolysis and diffusion through the cell wall, there is no need to provide a means of refilling, and the cell can be sealed with the exception of the pressure relief valve. This led manufacturers to describe the cells as 'sealed for life' and 'totally sealed'. This refers only to electrolyte loss and air ingress; It does not refer to gas release. As there is no free acid, the cells can in principle be operated in any orientation, although most are not recommended by manufacturers.

Many people, however, still believe that VRLA cells emit no hydrogen.

By virtue of their design, VRLA cells have a number of features which make them more suitable for new installations than the traditional Planté cell in certain circumstances :

> Increased kWh/m³,
> Less maintenance,
> Lower initial cost,
> Lower Hydrogen emmissions
> Improved seismic properties,
> Easier construction,
> Choice of Battery Orientation.

Ultimately, the main attraction to the customer is the fact that VRLA cells emit considerably less hydrogen than an equivalent Planté cell, and as a consequence, the ventilation requirement is greatly reduced. This eliminates the need for dedicated battery rooms, allowing the equipment to be installed closer to its need, even in cubicles alongside the charger equipment. This leads to substantial savings in cabling costs.

There are five main standards dealing with the use of Valve Regulated Lead-Acid cells. Of these, three quote a maximum hydrogen release rate for the cells, whilst one BS6745 (reference 1) intended for mobile applications does not, but requires that cells are marked 'Do not charge in a gas tight container'.

So what is the required ventilation for a battery of VRLA cells? Let us use a simple example to illustrate.

A backup DC power supply battery of 110V 48Ah, consisting of 54 cells is mounted in the base of a cubicle, that also houses the charging circuit. The total cubicle volume is 280 litres, of which the battery and charger equipment occupies 80 litres.

If a cell is said to conform to BS6290 pt 4 : 1987 (reference 2) the standard states that the maximum hydrogen release rate under normal charging conditions should be less than 5% of that expected from the direct electrolysis of water by the same current. A cell could be expected to draw a 50mA charge current per 100Ah of capacity. In the example quoted above, a release of about 30cc of Hydrogen per hour could be expected.

Meanwhile, the IEC (reference 3) and CENELEC (reference 4) draft standards quote the maximum release rate as being 10cc per cell per month per Ah. This results in a figure of 36cc per hour, not dramatically different.

The lower concentration limit for ignition of Hydrogen in air is 4%, with an ignition energy of 0.02mJ (Reference 5). Even if the cubicle described above were completely sealed, it would still take in excess of nine days to reach this value, although we would never wish it to reach anywhere near the concentration required for ignition. If the gas rate were to rise, then the time taken to reach ignition concentrations is greatly reduced.

The charging current and as a result the gassing rate increases with charger voltage, a process that is used in boost charging of a battery. Many manufacturers recommend setting the charger voltage to 2.40 Volts per cell from the normal float charge of 2.27 Volts per cell to achieve a fast recharge. A charger in boost condition can result in a current which can be as much as ten times normal float current.

The charger voltage can rise above normal float voltage for a variety of reasons such as an incorrectly set charger float voltage or a charger fault. The voltage of a VRLA cell drops with an increase in temperature. Consequently if the charger voltage is set when the ambient temperature is low, i.e. during the winter months, and the temperature subsequently rises, the voltage of the cell will reduce and the charging current increase leading to increased gassing.

If there is insufficient cooling of the battery, then an increase in charging current will lead to a warming of the battery which will in turn lead to an increased current. Over a period of days a runaway situation will develop causing extensive overheating of the battery, excessive Hydrogen gas emissions and ultimately cell destruction.

There are ways to minimise these effects ;

1. Avoid installing valve regulated cells in locations of high or variable ambient temperature.

2. Use a charger that automatically adjusts the charger voltage by measuring **battery** temperature, where temperatures may vary.

3. Provide good ventilation of the battery for cooling purposes. Such ventilation will assist the dispersal of hydrogen produced.

The IEC and CENELEC drafts quote a limit of gas release of six times that at float current with a charger voltage which is eight percent higher than float, i.e. ~2.4 Volts per cell. This means our example could gas at over 200cc/hr, and the compartment in the example above would reach ignition concentrations in just over a day and a half if it were completely sealed.

BS6133 gives the user guidance on ventilation of cubicles containing VRLA cells. It recommends that air should be passed through the cubicle to ensure that the concentration of hydrogen never exceeds 1%, thus providing an ample safety margin. If we take the example, quoted above. To keep the Hydrogen concentration below 1%, whilst only inducing one air change per hour, the battery could gas at over ten times the overcharge quoted by CENELEC and IEC.

This flow of one air change per hour can be achieved easily through a vent at the bottom of the cubicle and a raised top, as illustrated in Figure 2. In fact this arrangement would normally achieve many air changes per hour by natural convection if the cubicle is largely free of obstructions. If this cannot be achieved, the simple addition of a fan will provide ample ventilation, although a low flow-rate alarm would now be required to detect a failure of the fan. The use of a mechanical device is not, however, recommended due to the unreliability of such equipment.

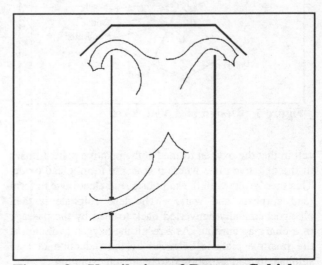

Figure 2 : Ventilation of Battery Cubicle

So is there a potential danger if we do not provide adequate ventilation with these cells? The answer is unfortunately yes. They may not produce a significant quantity of gas in comparison to flooded cells during normal trickle charging, but the danger is still there. There have been a number of reported serious incidents, one with fatal consequencies.

On that occasion a cubicle contained a nominal 110V battery of 48Ah capacity located in a battery compartment at the bottom, with the rectifier and other charger electronics being located in the top half of the cubicle. The unit had a free air volume of 200 litres, the same as the example quoted earlier. The charger cubicle was of conventional design containing relays, switches and cubicle heater.

The cubicle, along with another of a similar design was located on a hill-side, in the open air. An explosion of considerable force occurred within one cubicle, completely destroying it. The door and left side-panel were thrown a considerable distance, whilst the right side-panel was thrown against the adjacent hydraulic cubicle, where it was deformed to the shape of the compressed air cylinders. The contents of the cubicle, consisting of the rectifier/charger electronics and the battery were left as a tangled heap at the bottom of the cubicle.

The key feature of the design of this cubicle is that it had been specified to be weatherproof owing to its location outside, and subject to the elements. It was therefore substantially built to an IP55 rating. The doors, and all panels were sealed to prevent the ingress of rain or debris, resulting in no provision for ventilation. There had been a problem with the charger earlier following a thunderstorm, and it is believed that the charger voltage may have been faulty with a high charging voltage. The amount of hydrogen released was enough to allow an explosive concentration to be reached within a day. A small spark, maybe from a relay in the cubicle changing state, would have been enough to ignite the hydrogen. Small vents, designed to allow a passage of air but not rain, at the top and bottom of this cubicle, such as those described earlier, would certainly have prevented this tragedy.

Sadly this is not a unique incident. At a recent IEE discussion meeting on batteries, a designer was surprised by the response when he stated he wished to install a battery in a sealed container in the desert. He wanted it sealed since he was going to install it below the surface to reduce the temperature fluctuations between night and day. He questioned whether there would be any problems. We were happy to advise him.

In summary, it is important to recognise that valve regulated lead-acid cells do release hydrogen gas in normal use, although in relatively small quantities. It is when the charger is incorrectly set or has a fault that this release is increased significantly.

The cells are safe, if adequate ventilation is provided. The provision of ventilation to allow for cooling of the electronics and the battery should in most cases be sufficient to disperse the hydrogen released. It is essential that such ventilation is provided.

References :

1. British Standard "Portable Lead-Acid Cells and Batteries - Specification for Performance, Design and Construction of Valve Regulated Sealed Type" BS6745: Part 1: 1986

2. British Standard "Lead-Acid Stationary Cells and Batteries - Specification for Lead-Acid Valve Regulated Sealed Type" BS6290: Part 4: 1987

3. Draft IEC Publication " Stationary Lead-Acid Batteries; General Requirements and Test Methods; Part 2: Valve Regulated Types" Draft IEC 896-2 (1991)

4. Draft European Standard "Stationary Lead-Acid Batteries - General Requirements and Test Methods - Valve Regulated Types" CENELEC pr En50105: 1992

5. British Standard Code of Practice "Safe Operation of Lead Acid Stationary Batteries" BS 6133 : 1995

SYNCHRONOUS OR INDUCTION GENERATORS ? - THE CHOICE FOR SMALL SCALE GENERATION

D S Henderson

Napier University, UK.

INTRODUCTION

From the beginning of this decade, there has been a significant increase in the extent of small scale generation. This is occurring partly as a consequence of two Government targets, namely to have 5000 MW of combined heat and power (CHP) and 1500 MW of new renewable energy capacity in service by the year 2000. The latter is being implemented through the vehicle of the NFFO (Non-Fossil Fuel Obligation), and its equivalents in Scotland and Northern Ireland. More often than not, this generation is being connected to the distribution network and is termed *embedded generation*. The exceptions to this are mainly large CHP schemes which are outwith the scope of this paper.

Due to the combined pressures of the inherent low cost of induction generators, and the relatively high unit-cost (cost/kW) of small scale generation projects, there is a tendency to automatically specify induction generators. It is true that the induction generator offers a number of distinct technical advantages over the synchronous generator. These include simpler excitation, simpler starting and control requirements, robust construction and a relatively low contribution to fault levels. However their selection must be carefully considered, particularly to avoid potential problems with self-excitation.

This paper offers a timely review of the different operating characteristics of induction and synchronous type generators and highlights the various different technical and economic factors which must be considered when specifying and choosing the type of generators for small scale generation systems.

THE EXTENT OF NEW GENERATION

A study of the NFFO and SRO (Scottish Renewables Obligation) programmes gives a good example of the nature of this new generation. According to Bevan, (1), a total of 320 MW of declared net capacity (DNC) is now on-stream from the first two NFFO orders. This results from a total of 140 projects, giving an average DNC of 2.3 MW DNC per project. The contracted capacity of both the third order of the NFFO and the first SRO were larger that originally expected, with a total DNC in excess

TABLE 1 - The capacity of the SRO-1 and NFFO-3. Source - the DTI Review Supplement (2)

Technology Band	Contracted Capacity MW (DNC)	Number of projects
Wind generation	211.2	67
Hydro power	31.8	30
Landfill gas	82.1	42
Municipal and industrial waste	245.7	22
Energy crops & agricultural & forestry waste	132.7	10
Total	703.5	171

of 700 MW between them. The results of these orders are combined and given in Table 1.

Detailed analysis of the SRO determines that all of the projects have a DNC of less than 10 MW and that exactly 50% have a DNC of less than or equal to 2 MW. A similar result emerges from analysis of the NFFO-3; only 15 projects have a DNC in excess of 10 MW and approximately 50% (75 projects) have a DNC less than 2 MW. These figures amply demonstrate the extent of small scale generation which is intended to be connected to the system in the near future. The DTI, (2), expect that perhaps only 50 - 60 % of this capacity will proceed to the commissioning stage, nevertheless, there will still be a considerable increase in the level of embedded generation as a result of these, and future, orders.

Of equal importance is the geographical spread of the location of these projects. A glance at the maps issued with the NFFO-3 and SRO announcements show that the projects are scattered, quite literally, from Helmsdale in the North East of Scotland to Trannack Downs in Cornwall. Inherently, many of these renewable energy generation sites are in very remote situations, taking advantage of the natural resources e.g. the west coast for wind power installations and the mountains of Scotland and Wales for hydro power installations.

Opportunities and Advances in International Power Generation, 18–20th March 1996,
Conference Publication No. 419, © IEE, 1996

As the UK's grid network was established, the generation was usually performed in large power stations and the energy fed directly into the system at transmission level voltages. In other words the distribution network was not originally designed to cope with generation. Not surprisingly, in these remote locations, the distribution system is not necessarily capable of catering for the demands of embedded generation, not at least without some careful analysis and design, a point which some developers are prone to overlook.

The plant component which is most affected by this situation is the generator, and an informed choice must be made between the deployment of an induction machine or a synchronous machine. To do this, the engineer must be aware of the differences in operation between the two types of generator and their different interactions with the distribution system.

GENERATOR OPERATING CHARACTERISTICS

Type of Generator

Of the two principal types of generator available, only the synchronous generator is inherently suitable for operation on an electrically isolated system to supply the required voltage, frequency and power at the correct power factor. The induction generator cannot do this, it must be connected to a system having a substantial synchronous machine capacity, typically 90% according to Smith, (3). This is because the induction machine cannot produce its own excitation current to establish its magnetic field. Consequently it is of key importance to ensure that the system is capable of providing that excitation current.

Synchronous Generators

Excitation System. The main difference in the construction and operation between the two generator types is in the rotor construction and development of the rotor field. In the case of the synchronous generator, the magnetic field on the rotor is developed in discrete North and South poles usually through the action of passing a direct current through insulated rotor windings. Power to the rotor is fed via sliprings or by a brushless exciter system. The power is derived from the main generator terminals or from a permanent magnet generator driven from the generator shaft, so called self-excited schemes.

This is the basis on which the synchronous machine is capable of independent operation, however, clearly this design potentially carries a size, cost and reliability premium when compared with the induction machine.

Starting and Control. Whilst the synchronous machine is run up from standstill by the prime mover, the excitation supply must be made available. With the self excited schemes described above, this is not usually a problem. Unless the generator is intended to run in an electrically isolated situation, then synchronising equipment is necessary to ensure that the generator and the grid have the same voltage and frequency and that they are in-phase when the circuit breaker is closed. Although most small scale generating schemes are operated independently of the loading and voltage conditions on the grid, a few are installed to support the distribution system. Synchronous generators thus offer an advantage in certain embedded generation situations as independent control of both the power and the reactive power is possible, and by implication, so is control of the power factor. Again, there is a premium in the cost, size and reliability of the equipment necessary to perform these functions.

Fault Level Contribution. The fault current of a synchronous generator is likely to contribute to the system fault level, a factor which must be considered when examining the effect that the installation will have on the local network. Confirmation that the local system could withstand the increased fault level that would result is necessary.

Induction Generators

Excitation System. In the case of the induction machine, the rotor windings are usually un-insulated solid copper bars, shorted-out at each end of the rotor (cage construction). The field in the rotor winding is induced by the application of the rotating field. Hence in comparison to the synchronous generator, the rotor construction results in a smaller, cheaper and more rugged construction. The potential disadvantage in this case is that the supply system must provide the magnetising power.

Starting and Control. To start the induction generator, the set is brought up to speed by control of the prime mover. At, or slightly above, synchronous speed the circuit breaker is closed. The desired load condition is then set by the prime mover controls. The induction generator must be driven at a speed in excess of the synchronous speed, otherwise no power will be delivered. The operation of the induction generator is in the region of negative slip and is in the region of the circle diagram below that normally shown for a motor e.g. by Say, (4). The operating power factor is a function of the machine design parameters and the operating slip.

Unlike the induction motor design where the rotor resistance is optimised for starting purposes, the induction generator can be designed to have minimal rotor resistance and hence it will normally have a relatively high efficiency. There is an in-rush of current as the induction generator is initially connected to the grid. At the instant of starting, the magnitude of this current will be very similar to that obtained when direct-on-line starting an induction motor, (5 - 8 times full load current), however it usually lasts for only a few cycles. This in-rush current does not normally cause any problems unless, as stated by Allan, (5), the line or system to which the generator is connected is rated closely to the generator output in which case a severe voltage drop would occur. Where voltage drops would be problematic, alternative connection methods are possible, such as series resistor and "soft-start" thyristor control described by Parkkonen, (6).

When the supply system features auto-reclosure devices then the induction generator offers a further advantage as it does not require to be synchronised as such. However, extreme care must be taken in situations where there is the possibility that the induction generator is left connected to an excessive system capacitance on loss of the grid. In this situation, resonance and self-excitation can occur which could produce dangerously high voltage levels within the machine. This is a situation which is all the more likely to occur on smaller rated machines and a problem which is exacerbated by the local connection of power factor correction capacitors (5). Consequently the protection system must often include a protection relay to prevent this situation occurring, as directed by Engineering Recommendation G.59, (7), in the UK.

Fault Level Contribution. The contribution to the system fault level from the induction generator is low compared with that of the synchronous generator. Clearly this is another advantage in favour of the induction generator, particularly where the local system is already operating near its fault level capacity.

Induction Motors. The induction generator is in essence the same machine as the induction motor, only it is driven at speeds above synchronous speed rather than running at speeds just below. Understandably, therefore, there are many proponents of the application of standard induction motors to generating installations. In certain situations this may have its merits, however two factors must be noted. The first is that, as mentioned previously, the induction motor is usually designed with a significant value of rotor resistance for starting purposes, whereas in generator mode this can minimised. This condition thus adversely affects the efficiency of the machine when operating as a generator. The second factor to be considered is overspeed capability. Certain applications, particularly hydroelectric installations, may be subject to high overspeed conditions. A standard induction motor will not be designed to withstand an overspeed condition; a correctly designed induction generator would be.

ECONOMIC FACTORS

The per-unit capital cost (£/kW) of the installation is always an issue when considering the installation of any new generating plant. Although a figure of £2,000/kW is often used, e.g. by Freris, (8), for global calculations of the extent of the market, it is widely recognised and noted by Duckers, (9), that the target figure for economic success is less than £1,000/kW. As indicated in the foregoing sections, induction generators normally have a lower capital cost than synchronous machines as they have lesser excitation and control requirements and consequently they are smaller and lighter. This situation will also be reflected in the running costs, as the induction generator will have less maintenance and a lesser spares-holding requirement. Hence there are clear economic pressures in favour of the deployment of induction generators.

One exception to this is where the power rating overlaps with the highly competitive diesel-gen-set market, Jenkins, (10). Again a warning has to be given in situations where high overspeed conditions may prevail, as the standard generator designed for diesel application would not have any significant overspeed capability.

SUMMARY

Without doubt the UK is currently witnessing a significant increase in the number of small scale generation projects being installed. This is due partly as a result of the CHP and Renewable Energy programmes initiated by the Government.

There is usually a straight choice between the induction type and synchronous type of generator specified for these installations. For an electrically isolated system, the synchronous generator is the only machine which is inherently suitable, as the induction machine requires a supply of excitation current. This requirement means that the induction machine must be connected to a system which is known to be capable of providing that current.

When this condition has been satisfied, the induction generator offers several advantages over the synchronous machine; a simpler excitation system, a simpler starting and control system and a lower fault level contribution. These accrue to produce a system which is more rugged and usually has a lower capital cost than that for a synchronous machine. Warning must however be made in respect of potential problems for the induction generator associated with voltage-drop on starting and self-excitation on loss-of-grid.

Understandably, the Engineer is "encouraged" to specify an induction generator in every instance of embedded generation, however a conscious decision must always be made by the Engineer to confirm that it is indeed the correct technical choice.

ACKNOWLEDGEMENT

The author wishes to thank Napier University for the opportunity to prepare this paper.

REFERENCES

1. Bevan G, 1995, "The developing role of renewable energy sources", PEJ, Vol. 9, No 4, 154 - 155.

2. 1995, Review Supplement, Department of Trade and Industry.

3. Smith R B, 1985, "Mini hydro electric installations", BEAMA Technical Seminar for NEB -Thailand.

4. Say M G, 1983, "Alternating Current Machines", 275, 5th Edition, Pitman, London, UK.

5. Allan C L C, 1960, "Water-turbine driven induction generators", Proc. IEE, Part A, Vol. 107, No 36, 529-550.

6. Parkkonen M, 1993, "The connection of induction generator to the distribution electricity system", Proceedings of Hidroenergia '93, Section II, 39-47.

7. 1985, "Recommendations for the connection of private generating plant to the electricity boards' distribution systems", The Electricity Council Chief Engineers' Conference, Engineering Recommendation G.59.

8. Freris L, 1995, "Renewable energy", PEJ, Vol. 9, No 4, 159-163.

9. Duckers L, 1995, "Water power - wave, tidal and low-head hydro technologies", PEJ, Vol. 9, No 4, 164-172.

10. Jenkins N, 1995, "Embedded generation", PEJ, Vol. 9, No 3, 145- 150.

OFF-LINE UPS SYSTEM WITH OPTIMUM UTILIZATION OF POWER ELEMENTS

S. A. Hamed

University of Jordan, Jordan

Y. Al-Shiboul

Jordan Telecommunications Corporation, Jordan

ABSTRACT

In this paper, a new configuration of the off-line UPS system is presented. The system is designed with the least number of power components, which are the most expensive elements of the system, implying optimum utilization of these components. A simple booster winding is connected in series with the high voltage winding of the system transformer during the inversion mode to satisfy the load requirements. The control circuitry of the system is designed to facilitate this arrangement, and to ensure soft and synchronized transfer of the load from and to the mains. A systematic procedure is developed to select the ratings of the power elements of the system, including the booster winding. The availability of these elements, the sensitivity of the load to voltage fluctuations, and the control strategy of the inverter circuit are involved in this procedure. The proposed system was practically implemented, tested, and proved to operate satisfactorily under different loading conditions. It has the features of reduced weight, size, cost and improved reliability.

INTRODUCTION

Uninterruptible power supplies (UPSs) ensure continuous power flow to critical loads in the events of AC mains failure or disturbances. Several UPS technologies are currently available. These include ferroresonant, pulse width modulation (PWM), hybrid and rotary UPSs [1]. Until recently, PWM based units are the most common of these. Off-line and on-line UPSs are the major types of UPS systems. The choice between one type or the other is mainly affected by the cost, reliability, efficiency, and the technical performance characteristics of the system. Off-line systems are generally simpler, smaller, and cheaper. Figure 1 shows the basic structure of the conventional off-line UPS system, where both the battery charger and the inversion circuits have their own power elements. Two ordinary power transformers and two sets of power diodes (4 elements each) are required. Since the rectifier and the inverter are never simultaneously participating in the process of power flow into the load, the utilization of the power elements of this configuration, which are the most expensive elements of the system, is relatively low. Great efforts are being put forth nowadays to implement small and economic UPS systems. One critical area that is useful in this context, is the magnetic structure of the system. Magnetic elements are present in every UPS system. UPS configurations with triport transformers and

external non-integrated stabilizing coils are recently employed. An off-line UPS system, that was built over a triport transformer and needs no stabilizing coil,was reported by Martinez et al [2]. Aoki et al [3] proposed a UPS system that does not use a conventional separate battery charger.

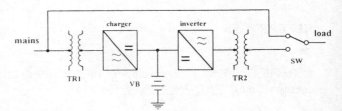

Fig. 1: Construction of the Conventional Off-Line UPS.

This paper presents a new configuration of the off-line UPS system. The system has the features of reduced weight, size and cost when compared with the conventional UPSs. In the following sections, the power circuit configuration, the control circuitry and the principle of operation of the system are described. Also, a systematic procedure is developed for proper selection of the power elements of the system in order to meet the requirements of the load and to suit the inverter control strategy.

POWER CIRCUIT CONFIGURATION OF THE PROPOSED SYSTEM

Figure 2 shows the power circuit configuration of the proposed off-line UPS system.

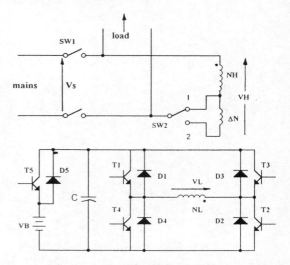

Fig. 2: Power Circuit Configurations of the Proposed System

Opportunities and Advances in International Power Generation, 18–20th March 1996, Conference Publication No. 419, © IEE, 1996

This circuit comprises the following power elements:

- One common two-winding power transformer N_H-N_L. This transformer serves as a step-down transformer during the *normal mode* of operation, with a transformation ratio $A_n = N_L/N_H = V_L/V_S$, where V_S is the rms value of the mains nominal voltage, V_L is the rms voltage of the low-voltage side of the transformer, N_L and N_H are, respectively, the number of turns of the low-voltage and high-voltage sides of the transformer.

During the *inversion mode*, the booster winding, which has a number of turns ΔN and requires no additional magnetic structure, is connected in series with the high-voltage winding of the common transformer in order to satisfy the load voltage requirements. Here, the transformer serves as a step-up transformer, with a transformation ratio $A_i = N_L / (N_H + \Delta N) = V_{L1} / V_{H1}$, where V_{L1} and V_{H1} are, respectively, the rms values of the fundamental components of the inverter output voltage and the load terminal voltage. The number of turns of the booster winding ΔN is mainly affected by the control strategy of the inverter circuit.

- One set of power diodes D1-D4, which performs as a bridge rectifier, through which the battery is charged during the *normal mode*. They also serve as freewheeling-feedback diodes during the *inversion mode*.

- Transistors T1-T4, which are the power switching elements of the inverter circuit. These are driven to generate a sinusoidally pulse width modulated voltage.

- Transistor T5, which is driven to operate in the linear region during the *normal mode*, to serve as a regulator that satisfies the voltage and current limitations of the battery. During the *inversion mode*, this transistor is driven into saturation in order to complete the path for the feedback currents.

- Diode D5, added to complete the forward current path during the *inversion mode*.

- Capacitor C, which serves as a smoothing capacitor of the charging voltage. It also serves as a snubber that protects the inverter circuit against any over voltages.

- Line interrupter SW1, serves as a bi-stable switch that is closed during the *normal mode*, and opened in the event of mains failure. The change-over switch SW2 is controlled to be at position 1 during the *normal mode* to isolate the booster winding ΔN, and at position 2 during the *inversion mode*. Both switches are controlled by the same signal, and, practically, can be the contacts of the same relay.

Figure 3 shows the power circuit configuration of the charging circuit, from which it may be realized that only the power elements of the inverter unit (not all of them)

are involved in the charging process. Hence, no separate charging unit is required by the system.

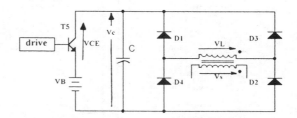

Fig. 3: Power Circuit Configuration of the Charging Unit.

PRINCIPLE OF OPERATION

During the *normal mode* of operation, the load is fed from the mains through the contacts of the line interrupter SW1, and the battery is kept fully charged. When the mains fails, the line interrupter SW1 opens to disconnect the AC mains, and the change-over switch SW2 changes from position 1 into position 2. The booster winding ΔN is then connected in series with N_H to participate in the voltage transformation during the inversion process. The inverter continues to supply the load almost simultaneously by applying the appropriate control signals to the bases of transistors T1-T4. This is arranged by the inverter control circuit, which is kept ready in a hot stand-by mode. During the *inversion mode*, the diode D5 provides a path for the current of the inverter transistors, and transistor T5 is driven into saturation in order to complete the feedback current path. In case of AC mains restoration, the inverter is resynchronized with the mains such that when the inverter output voltage is in phase with the mains voltage, the inverter stops and the line interrupter SW1 is closed to feed the load from the AC mains, and SW2 changes to position 2. This assures soft and synchronized transfer of the load to the AC mains. The charging unit of Fig. 3 will then begin to recharge the battery to be ready for the coming AC mains failure. The whole process is supervised and controlled by the UPS control circuitry, which is described by the block diagram of Fig. 4, and comprises the following:

(A) Charging Control Circuitry: Block (1) of Fig. 4 represents the control circuitry of the charging unit. This unit drives the regulation transistor T5 to operate in the linear region during the *normal mode* of operation. It comprises elements that sense the two limiting states of the battery (i.e. battery fully-charged and battery end of discharge), in order to control the charging current and to limit the battery voltage such that it does not exceed the maximum charging voltage $V_B(\text{max})$.

(B) Inverter Control Circuitry: Different control strategies can be employed to reduce the harmonic contents of the inverter output voltage during the *inversion mode*. Sinusoidal pulse width (SPWM) is

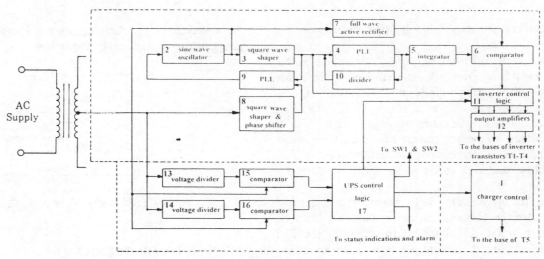

Fig. 4: Block Diagram of the Control Circuitry.

employed since it is the most common modulation strategy of these. The waveform of the output voltage with a unity modulation index is shown in Fig. 5.

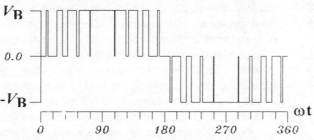

Fig. 5: Inverter Output Voltage Waveform

The control circuitry of the inverter comprises blocks (2)-(12). The sine oscillator (2), in combination with the square wave generators (3) and (8) and the phase locked loop (9), generates a sine wave that is varying at the mains frequency and synchronized with the AC mains voltage. In case of mains absence, this oscillator continues producing the sine wave signal at its output. The output of the oscillator is then used to generate the triangular wave at the required modulating frequency f_c by means of the phase locked loop (4), in combination with the divider (10) and the integrator (5). The output of the oscillator (2) is rectified using a full-wave active rectifier (7) and then compared with the triangular wave by the comparator (6) in order to generate the appropriate drive signals of the inverter transistors. The output of the comparator (6) is then fed to the inverter control logic (11). The function of block (11) is to start and stop the operation of the inverter, and to drive the inverter transistors. Block (12) represents the power amplifiers of the drive signals of the inverter transistors.

(C) UPS Control Circuitry: This is the circuit that supervises and controls the overall operation of the UPS system. In case of mains presence, the mains voltage V_s is stepped down and compared with two adjustable reference voltages (blocks 13-16). The two reference voltages are set according to the load sensitivity to voltage

fluctuations, usually given by the Power Tolerance Envelope (PTE) of the load [4]. When the voltage of the mains is within the two reference values, normal operation is maintained. Also, the synchronizing signal is supplied to the circuit which controls the operation of the inverter to ensure a synchronized running of the sine wave oscillator with the AC mains. In case of any deviation, signals are initiated by block (17) to switch off the line interrupter SW1, change SW2 into position 2, start the inverter, drive T5 into saturation. The synchronization signal is also interrupted to allow free-running of the sine wave generator. If the deviation of the mains voltage lasts less than a certain period of time, that is usually acceptable by the PTE curve of the load, no actions will take place by the control circuit. Appropriate indications and alarm signals for close monitoring of the system operation is also initiated by this circuit.

SELECTION OF THE POWER ELEMENTS

A systematic procedure is developed to select the ratings of the main power elements of the proposed system. Details of this procedure are not included due to space restrictions. This selection, however, is made as follows:

From the charging circuitry, the nominal battery voltage V_{BN} is given by the expression:

$$V_{BN} = [\sqrt{2} A_n / K_1] V_S \qquad (1)$$

where K_1 is a coefficient defined as the ratio between the maximum charging voltage $V_B(max)$ and the nominal battery voltage (i.e., $K_1 = V_B(max)/V_{BN}$). For a given nominal voltage V_S, V_{BN} is chosen according to the above expression, taking into account the availability of the battery, the power transformer, and the economic considerations.

At the beginning of the discharge process, the battery voltage drops promptly from $V_B(max)$ to V_{BN}. V_B will then expected to vary within the range $V_B(min) \le V_B \le V_{BN}$ during the inversion mode. As a result, the fundamental terminal voltage varies in the range $V_{H1}(min) \leq V_{H1} \leq V_{H1}(max)$. The two limits of V_{H1} are found to be:

$$V_{H1}(max) = \sqrt{2}\, \frac{A_C}{A_i}\, \frac{K_3}{K_1}\, V_S \qquad (2)$$

$$V_{H1}(min) = \sqrt{2}\, \frac{A_C}{A_i}\, \frac{K_2 K_3}{K_1}\, V_S \qquad (3)$$

where: K_2 is the ratio of the minimum allowed battery voltage $V_B(min)$ and the nominal battery voltage (i.e., $K_2 = V_B(min)/V_{BN}$).

K_3 is a coefficient depends upon the control strategy of the inverter circuit and the modulating frequency. It is defined as the ratio between the rms value of the fundamental component of the inverter output voltage during the *inversion mode*, V_{L1}, and the battery voltage V_B (i.e., $K_3 = V_{L1}/V_B$).

Based on the first limit above, the number of turns of the booster winding ΔN is determined and given as:

$$\Delta N = [(K_1/\sqrt{2}\,K_3) - 1]\, N_H \qquad (4)$$

From the load point of view, the UPS system should meet the sensitivity requirements of the load to voltage fluctuations, which can be reflected by the coefficient K_4, where $K_4 = V_{H1}/V_S$. The safe range of K_4 is usually given by the load PTE curve as an indication to how critical the load is. For instance, the PTE curve of the computer load shows that under steady state operation, coefficient K_4 is allowed to vary within the range $0.87 \leq K_4 \leq 1.06$ [4]. Based on the second limit of (3) above, the relationship:

$$\frac{V_{H1}(min)}{V_S} = K_2 \geq K_4(min) \qquad (5)$$

should be satisfied when selecting the battery of the system.

A *250 VA, 220 V, 50 Hz* off-line UPS system was designed, implemented and practically tested using a running computer terminal as a load. The ratings of the power elements are:

- Battery ratings: $V_{BN} = 36\ V$, $V_B(max) = 39\ V$, $V_B(min) = 33.5\ V$. The corresponding battery coefficients are $K_1 = 1.0833$ and $K_2 = 0.9305$.
- Power transformer ratings: *300 VA, 30/240 V and 50 Hz*.
- Coefficient $K_3 = 0.707$, corresponding to a modulating frequency of *1 KHz*.
- Booster winding $\Delta N = 0.083\ N_H$.

During the *inversion mode*, the fundamental component of the load terminal voltage varies from *100 % V_S*, at the beginning of the discharge process, and *93.05 % V_S*,

when the battery is fully discharged. This satisfies the requirements of most load types. The interruption time during the transfer operation from and to the mains is shown to be short enough not to disturb the normal operation of the running computer which is classified as a critical load.

CONCLUSIONS

A new configuration of the off-line UPS system that requires the least power components is described in the paper. The system needs no separate battery charger since the power elements of the inverter circuit are arranged to accomplish the charging process during the *normal mode*. Only one common two-winding transformer and one set of power diodes (4 elements) are employed in the proposed system. From the power circuit point of view, this is made possible by inserting a simple booster winding in series with the high-voltage winding of the common transformer, connected during the *inversion mode* only. The control circuitry of the system was designed to ensure proper supervision and control of the system operation. A systematic procedure is developed to select the ratings of the back-up battery, the power transformer and the number of turns of the booster winding. This procedure takes into consideration the availability of these elements, the sensitivity of the load to voltage fluctuations, and the control strategy of the inverter circuit. Sinusoidal pulse width modulation (SPWM) control strategy is employed by the inverter stage in order to eliminate or reduce the amplitudes of the lower-order harmonics and to simplify the filtration requirements of the higher-order harmonics of the output voltage. A 250 VA model of the proposed system was practically implemented, tested and proved to operate successfully under different loading conditions. The interruption time during transfer operations was found to be short enough not to affect the normal operation of different critical loads such as a running computer. The utilization of the most expensive elements of the proposed system is relatively high, leading to reduced weight, size, cost and improved reliability, when compared with the conventional systems.

REFERENCES

[1] Albright A, 1992, "UPS System for Critical Applications", Proceedings of the Rural Electric Power Conf., pp. 1-9.
[2] Martinez S, Castro M, Antoranz R and Aldana F, 1989, "Off-Line Uninterruptible Power Supply with Zero Transfer Time Using Integrated Magnetics", IEEE Trans. Ind. Electronics Vol. 36, No.3, pp. 441-445.
[3] Aoki T, Yotsumoto K, Muroyama S and Kenmochi Y, 1990, "A New Uninterruptible Power Supply with Bidirectional Cycloconverter", Proceedings of the International Telecoms. Energy Conf. (INTELEC), pp. 424-429.
[4] Joshi H, 1991, "Consideration for Selecting and Sizing of UPS system for Critical Loads", Proceedings of the Intern. Telecoms. Energy Conference, pp. 453-458.

SYNCHRONOUS MACHINE PARAMETERS, THEIR INFLUENCE ON THE A.C. VOLTAGE DISTORTION OF ISOLATED SYSTEMS

S. Salman   J. Penman   K.S. Smith   I.D. Stewart

University of Aberdeen, U.K.   B.P. Exploration, U.K.

ABSTRACT

In general the power system of an offshore platform consists of synchronous generators, transformers, converters and large induction motors. The converters can be a significant part of the total system load and the system voltage waveshape is therefore distorted. This distortion is due to the non-linear nature of the converter thyristors.

The aim of this paper is to relate the parameters of a single synchronous generator to the a.c. voltage distortion at the machine terminals on isolated systems. This in turn will demonstrate the possibility of providing measurement techniques using the voltage and current waveforms for estimating the parameters of a synchronous generator.

INTRODUCTION

A knowledge of synchronous machine parameters is essential for the modelling and analysis of electrical machines. For offshore power systems for example, machine parameters are needed for fault level and transient stability studies which are carried out on offshore systems to ensure the safe operation of the platform.

The parameters of synchronous generators are estimated using tests which are in accordance with British Standards (1). However for any particular generator the tests can produce different values for the same parameter. This poses a problem as to which of these test values to use in fault level calculations and stability studies.

The aim of this paper is to relate the parameters of a single synchronous generator to the a.c. voltage distortion at the machine terminals. This will go some way towards understanding the more general case where one or more generators are connected to the same busbar. The possibility of providing measurement techniques using the voltage and current waveforms for estimating the parameters of a synchronous generator is also described.

The system studied here is a simplified model consisting of a generator connected to a three phase reactor, a thyristor bridge and a d.c. load as shown in

figure (1). Due to the presence of a thyristor bridge (non-linear load) current will flow in the damper windings, i.e. the generator is operating continuously in the transient state. The voltage waveshape which is produced is therefore a function of both the steady state and transient parameters.

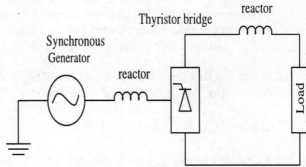

Figure 1: A model of a synchronous salient pole generator connected to a thyristor bridge.

The following assumptions have been made in this paper,

The magnetic circuit is assumed linear, (i.e. magnetic saturation is neglected), the mathematical model and the associated programming are assumed correct, the direction of positive currents is out of the terminals for generator action, hence for the positive direction of the magnetic axes, negative flux linkages result due to positive stator current. Only one damper winding on the quadrature axis is assumed, i.e. $L_{kq1} = L_{kq2}$ and $r_{kq1} = r_{kq2}$.

ESTIMATING PARAMETERS OF THE SYNCHRONOUS GENERATOR

The tests used to estimate the steady-state and transient parameters of a salient pole synchronous generator with a rating of 2.75kVA, 130V, 50Hz, are generally those in accordance with British Standards. These tests include the no-load test, sustained three phase short circuit test, sudden three phase short circuit test, zero power factor test, the slip test, the d.c. decay test and estimation of the effective turns ratio, i.e. the turns ratio between the armature winding and the field winding.

The no-load test and the sustained three phase short circuit test are both used in the calculation of the direct-

Opportunities and Advances in International Power Generation, 18–20th March 1996,
Conference Publication No. 419, © IEE, 1996

axis synchronous reactance (X_d) as well as the armature resistance (r_s).

The sudden three phase short circuit test is used to determine the transient parameters such as the direct-axis subtransient reactance and time constant (X_d'' and T_d''), and the direct-axis transient reactance and time constant (X_d' and T_d').

The armature leakage reactance (X_{ls}) is determined from the zero power factor test and the method proposed by El-Serfi and Wu (2).

A slip test is used in the calculation of the quadrature- and direct- axes synchronous reactances (X_q and X_d).

The D.C. decay test which has been categorised by the British Standard as an 'unconfirmed test method', is used to determine the transient and subtransient reactances and their corresponding time constants for both the direct-axis and quadrature-axis (X_q', X_q'', T_q', T_q'').

The effective turns ratio is also required in the calculation of the armature leakage reactance (X_{ls}) and in the computer simulation, which is performed using the commercial package Saber (3).

MODELLING OF THE SYNCHRONOUS GENERATOR

In order to relate the parameters of a single synchronous generator to the a.c. voltage distortion at the machine terminals on isolated systems, the model represented by figure (1) was set up both in the laboratory and simulated using Saber.

The generator is represented in Saber in terms of voltage equations, (Krause (4)), these equations model both the stator and the rotor. The stator windings are the three phase armature windings with a leakage inductance (L_{ls}) and resistance (r_s), and the rotor is represented by a field winding and three damper windings, kd, kq1 and kq2. The field winding is represented by a field leakage inductance (L_{fd}) and a field resistance (r_{fd}), while the damper windings are expressed in terms of inductances and resistances, e.g. the damper winding kd is modelled in terms of L_{kd} and r_{kd}.

The generator parameters (inductances and resistances) used in Saber were estimated using the tests described above. These tests were carried out on the laboratory generator.

The measured parameters together with the field voltage are used in Saber to determine the armature and damper currents. The product of the inductances and currents gives flux linkage, and by differentiating the flux linkage the voltage can be determined, e.g.

$$V_{sa} = -r_s i_{sa} + \frac{d\Psi_{sa}}{dt} \qquad (1)$$

The accuracy of the parameters is therefore essential for accurate calculation of the voltage waveshape.

To ensure the measured parameters are correct a simulation was performed. The voltage waveshape produced was then compared with the voltage measured in the laboratory, Salman et al (5). Figure (2) shows that the two waveforms are very similar, i.e. the parameters used in Saber can be assumed to be of the correct value.

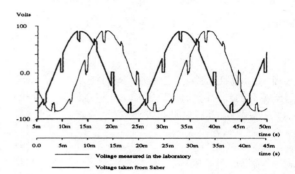

Figure 2: The voltage measured in the laboratory compared with the voltage obtained from Saber.

STUDYING THE EFFECT OF THE MACHINE PARAMETERS ON THE VOLTAGE WAVEFORM

Provided the d.c. load current, thyristor firing angle, generator speed and field current are kept constant, the voltage waveshape produced from the model of Figure (1) will be a function of the generator parameters (inductances and resistances). Hence to study the effect these parameters have on the voltage waveform, each parameter was altered by a certain percentage. A simulation was then carried out and the voltage waveshape produced was compared with the voltage obtained from the original unchanged model.

The synchronous generator parameters studied in this paper are the damper winding leakage inductances and their corresponding resistances (L_{kd}, L_{kq1}, r_{kd}, r_{kq1}), the stator leakage inductance and resistance (L_{ls}, r_s) and the magnetising inductance on both the direct- and quadrature- axes (L_{md}, L_{mq}). These parameter were calculated from the direct- and quadrature- axes parameters which were measured on the laboratory generator using the tests described above.

For example, the direct-axis magnetising inductance L_{md} is calculated from

$$L_{md} = (X_d - X_{ls}) * (Z_b / \omega_b) \qquad (2)$$

where X_d is the direct-axis synchronous reactance, X_{ls} is the stator leakage reactance, Z_b is the base impedance of the generator and

$$\omega_b = 2 * \pi * \text{frequency} \qquad (3)$$

The effect of varying the direct-axis (L_{md}) and quadrature-axis (L_{mq}) magnetising inductances

The study showed that when L_{md} was reduced by 35% of its original value, the result was a reduction in the voltage amplitude as shown in figure (3).

This is because the voltage is expressed as follows,

$$V_{sa} = -r_s i_{sa} + \frac{d}{dt}(\Psi_{sa\text{-}sa} + \Psi_{sa\text{-}sb} + \Psi_{sa\text{-}sc} + \Psi_{sa\text{-}kq} + \Psi_{sa\text{-}fd} + \Psi_{sa\text{-}kd}) \quad (4)$$

where
$$\Psi_{sa\text{-}sa} = -(L_{ls} + L_a - L_b*\cos(2*\theta_r))*i_{sa} \quad (5)$$
$$\Psi_{sa\text{-}sb} = -(-1/2*L_a - L_b*\cos(2*\theta_r - 2\pi/3))*i_{sb} \quad (6)$$
$$\Psi_{sa\text{-}sc} = -(-1/2*L_a - L_b*\cos(2*\theta_r + 2\pi/3))*i_{sc} \quad (7)$$
$$\Psi_{sa\text{-}kq} = L_{mq}*\cos(\theta_r)*i_{kq} \quad (8)$$
$$\Psi_{sa\text{-}fd} = L_{md}*\sin(\theta_r)*i_{fd} \quad (9)$$
$$\Psi_{sa\text{-}kd} = L_{md}*\sin(\theta_r)*i_{kd} \quad (10)$$
$$L_a = 1/3(L_{mq} + L_{md}) \quad (11)$$
$$L_b = 1/3(L_{md} - L_{mq}) \quad (12)$$

From equation (4) it can be shown that if L_{md} was reduced, then the total flux linkage is reduced. This causes a reduction in the value of the voltage, and when this occurs, the calculated current is also reduced.

However when L_{mq} was reduced by 35% of its original value, the change observed on the voltage waveform was a reduction in the depth of the commutation notch by 5%; the notch is caused by the short circuit which occurs when three thyristors conduct at any one time. A reduction of 4% in the width of the notch was also observed as shown in figure (4).

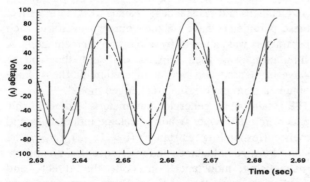

Figure (3): The effect of reducing L_{md} (dotted line) on the voltage waveform.

This reduction in the width of the notch indicates that the commutating inductance is a function of the parameter L_{mq}. This can be described by studying the system during commutation. Lander (6), shows that the commutation of thyristor one to thyristor three can be expressed in the following voltage equation as,

$$V_b - V_a = 2L \, di_c/dt \quad (13)$$

where $V_b - V_a$ can be derived by substituting equation (4) for phase a and phase b. The currents I_a, I_b and I_c in equation (13) take the following form during commutation,
$I_a = I_d - i_c$, $I_b = i_c$ and $I_c = -I_d$, where i_c is the commutation current, I_d is the load current and L is the commutating inductance of the model.

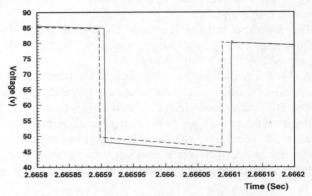

Figure 4: The effect of reducing L_{mq} (dotted line) on the terminal a.c. voltage.

When the parameter L_{mq} was reduced, a reduction in the commutation notch was noticed, i.e. the time dt was reduced. This resulted in the function di/dt being increased, the reduction in V_a and V_b however was insignificant, hence the value of the term $(V_b - V_a)$ remains the same. In order to balance equation (13), the value of L must be reduced, which indicates that L is a function of L_{mq}.

The effect of varying L_{ls} and the damper windings

The effect of reducing the stator leakage inductance (L_{ls}) by 35% was a reduction in the width of the commutation notch by 6%. This reduction can be explained in a similar way to the reduction noticed when the parameter L_{mq} was reduced, i.e. the commutation inductance (L) must also be a function of L_{ls}.

A study of the effect that the damper winding inductances and resistances have on the voltage waveshape was also carried out. Although these parameters are not a function of the voltage equation, they are used in the Saber simulator to calculate the damper winding currents which are needed in the calculation of the terminal a.c. voltages. For example the current in the damper winding kd can be determined from the following equation,

$$V_{kd} = r_{kd} i_{kd} + d/dt (\Psi_{kd\text{-}sa} + \Psi_{kd\text{-}sb} + \Psi_{kd\text{-}sc} + \Psi_{kd\text{-}fd} + \Psi_{kd\text{-}kd}) \quad (14)$$

where
$$\Psi_{kd\text{-}sa} = -(2/3)*(L_{md}*\sin(\theta_r))*i_{sa} \quad (15)$$
$$\Psi_{kd\text{-}sb} = -(2/3)*(L_{md}*\sin(\theta_r - 2\pi/3))*i_{sb} \quad (16)$$
$$\Psi_{kd\text{-}sc} = -(2/3)*(L_{md}*\sin(\theta_r + 2\pi/3))*i_{sc} \quad (17)$$
$$\Psi_{kd\text{-}fd} = L_{md}*i_{fd} \quad (18)$$
$$\Psi_{kd\text{-}kd} = (L_{kd} + L_{md})*i_{kd} \quad (19)$$

Hence when the inductance L_{kd} was reduced by 35% of its original value, the result was a decrease in the value of the damper current (i_{kd}). The reduction in the damper current (i_{kd}) had the effect of decreasing the width and increasing the depth of the commutation notch on the voltage waveform. A slight increase in the armature current by 0.02% was also observed. However when the damper winding inductance (L_{kq}) was reduced by 35%, the effect was an increase in the width of the commutation notch by 2%. The effect of reducing the field inductance (L_{fd}) by 35%, was a slight increase in the depth of the commutation notch.

The damper resistances for both the direct- and quadrature- axes (r_{kd} and r_{kq} respectively) had no effect on the a.c. voltage and current when they were reduced by 35%. A reduction of 35% in the field resistance (r_{fd}) however produced an increase in the amplitude of the a.c. voltage by 4%. This increase is due to the fact that a reduction in the field resistance resulted in an increase in the calculated field current (I_{fd}), which in turn gave rise to an increase in the calculated a.c. voltage.

The effect of varying the field resistance (r_{fd}) on the voltage waveform is greater than that noted when varying the field inductance (L_{fd}). This is due to the fact that a generator with a small rating will have a high resistance, hence the reduction in the resistance will have a more significant effect on the field current, which in effect will vary the terminal a.c. voltage produced.

Finally the armature resistance (r_s) was reduced by 35% and the effect was a small increase in the amplitude of the a.c. voltage and armature current.

This study has demonstrated the effect the generator parameters have on the voltage waveform. For example, the damper winding parameters effected the commutation notch of the voltage waveform, which in turn indicated that the voltage waveform is a function of the damper winding parameters. A similar pattern has been shown for variations in the other parameters of the synchronous generator.

Hence the possibility of estimating the parameters of the synchronous generator directly from the line voltage is achievable provided the following assumptions are considered;

The voltage waveform during steady-state operation is a function of the steady state parameters only, i.e. currents do not flow in the damper windings, and during transient operation (when a non-linear load such a thyristor bridge is connected to the generator) the voltage waveform is a function of both the steady-state and transient parameters.

The steady-state parameters are the direct-axis magnetising reactance (X_{md}), quadrature-axis magnetising reactance (X_{mq}) and the stator leakage reactance (X_{ls}). These parameters can be estimated directly from the line voltage when the generator is driven at no-load, and with a linear load.

The transient parameters which include the damper winding resistances and reactances for both the direct- and quadrature- axes and the field winding resistance and reactance can be determined from the line voltage and armature current when the generator is connected to a non-linear load such as a thyristor bridge.

CONCLUSION

The parameters of a synchronous generator are needed for fault level and transient stability studies particularly on offshore platforms.

The effect the generator parameters have on the voltage waveform has been illustrated, and it can be concluded that the steady-state and transient parameters are a function of the voltage waveform. Hence a method can be developed to estimate the parameters of the synchronous generator directly from the line voltage and current waveforms. This will be both faster and less expensive then the conventional methods.

ACKNOWLEDGEMENTS

The authors acknowledge the support of the Engineering and Physical Sciences Research Council, the Marine Technology Directorate and B.P. Exploration in funding the research work.

REFERENCES

1. British Standards, 1988, "General requirements for rotating machines", Part 104. Methods of test for determining synchronous machine quantities, BS 4999, 1988

2. El-Serfi A.M., Wu J., 1991, "A new method for determining the armature leakage reactance of synchronous machines", IEEE Transaction on Energy Conversion, Vol 6, No. 1, pp 120

3. Analogy, 1993, "Guide to writing templates" Analogy Inc, Release 3.2

4. Krause P.C., 1986, "Analysis of electric machinery", McGraw Hill, pp271-282

5. Salman S., Smith K.S., Stewart I.D., 1995, "Measurement of synchronous machine parameters; a practical experience", UPEC95, Vol 2, pp 785

6. Lander C.W., 1993, "Power Electronics", McGraw Hill, 3rd edition

A SOLID-STATE SYNCHRONOUS VOLTAGE SOURCE WITH LOW HARMONIC DISTORTION

Z Chen, E Spooner

University of Durham, UK

ABSTRACT

The power conversion requirements for a variable-speed permanent-magnet generator feeding into the network are discussed. A system using a rectifier, dc link, and grid-synchronised voltage-source inverter to process the power from a variable-speed, modular PM generator in a wind turbine is described The inverter phase angle can be adjusted continuously to control power and turbine speed to maximum energy capture from the wind and the output voltage amplitude can be adjusted to control reactive power.

Several harmonic elimination strategies are considered and the multi-pulse inverter is shown to be a cost effective solution with low current harmonics.

Experimental results from a laboratory model are presented along with computer simulation results which are in good agreement.

INTRODUCTION

Renewable energy is now a commercial reality in the UK. Wind, the least-cost source, is established as a viable means of generating electricity[1]. Wind energy conversion technology is developing towards direct-drive, variable-speed systems which increased energy capture, reduced mechanical stresses and aerodynamic noise. To connect the variable-speed wind energy conversion system to ac network, an interface is required. An ideal interface would optimise the operation of both the wind turbine and the electric power network.

The rapid development of semiconductor devices is extending the range of applications for power electronics in the electrical power system. For the last 30 years, applications have been primarily in high voltage direct current (HVDC) transmission but now a new generation of power electronic converters applied to reactive power compensators, active harmonic filters and other devices has brought about the concept of the Flexible AC transmission System[2] (FACTS). FACTS devices are compact and can increase the safe capacity of the ac network leading to reduced costs of transmission.

The paper presents a solid-state synchronous voltage source which acts as an ASVC (advanced Var Compensator) and which supplies real power to the network from the dc link of a variable-speed wind turbine generator. The inverter phase angle can be adjusted continuously to ensure maximum power capture from the wind and the output amplitude can be adjusted to control reactive power.

WIND ENERGY CONVERSION

System description

A direct-drive, variable-speed wind energy conversion system is shown schematically in Fig. 1.

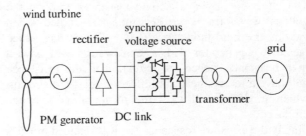

Figure 1 Variable speed wind energy conversion system

A PM generator is directly connected to the wind turbine and its output varies in frequency and voltage. The output is rectified to form a dc link and a voltage source inverter converts the power to ac for the grid. The inverter should be force-commutated with a high effective switching frequency to provide reactive power control and low harmonic distortion of the grid current.

Modular PM generator

The proposed power conversion system has been investigated in the course of studies to develop a direct-drive variable-speed PM generator[3,4,5,6]. Each module of the stator carries a single coil which provides a single-phase ac output.

The stator coils taken together form a multi-phase output, which after rectification, establishes a dc link with very little ripple because of the large number of phases. A dc link filter may not be needed. Diodes may therefore be used in the first conversion statge, ac to dc, for simplicity and economy. Ac capacitors may be used to enhance the power capability of the generator[6].

The diode AC/DC converter provides no direct means of controlling power. Instead the dc link voltage is used as the control variable for setting the power extracted from the generator. The dc voltage is controlled via the grid interface inverter which may be either line or force commutated. The force commutated synchronous voltage source inverter is discussed in detail below.

INVERTER STAGE

Inverter configuration

The elements of a suitable power circuit for a phase and magnitude controllable synchronous VSI are presented in Fig. 2. If the switches are self commutated, then they may be operated in a PWM pattern to produce an output at fundamental frequency with the desired magnitude and phase with respect to the grid voltage. Alternatively, the three-phase bridge may be switched in a six-step sequence and the phase adjusted with respect to the grid; the output magnitude being controlled by the dc chopper (S_D)

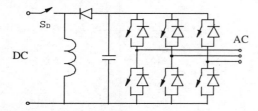

Figure 2 Schematic of voltage source inverter

Power relationships

Sinusoidal, three-phase balanced voltages are assumed and are depicted in the equivalent circuit and phasor diagram of Fig. 3. L_s represents the inductance between the output of the inverter ($V_{I(1)}$) and reference point of the ac system (V_s). Resistance is ignored. The inverter and inductor together mimic a synchronous generator, therefore the power flow can be expressed as:

$$P_s = P_I = \frac{V_s V_{I(1)}}{X_s} \sin \delta \qquad (1)$$

$$Q_s = \frac{V_s V_{I(1)}}{X_s} \cos \delta - \frac{V_s^2}{X_s}$$

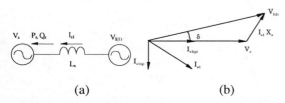

| (a) | (b) |

Figure 3 (a) Inverter-grid equivalent circuit
(b) Phasor diagram

Clearly, within limits set by V_s and L_s, any desired magnitude and direction of real and reactive power can be achieved by control of $V_{I(1)}$ and δ.

Real power control

For maximum energy capture, the shaft speed should be controlled so that the tip speed ratio (blade tip speed /wind speed) is held at the opimum value as wind speed varies until the rated power is reached. Subsequently, the power will be limited either by blade pitch angle control or by aerodynamic stall, yaw limiting is also being used in some turbines. Then the turbine will operate at the rated power until cut out speed is reached.

Speed can be controlled by the generator reaction torque which, in turn, is determined by the voltage magnitude and the power angle at the grid inverter. These are thus the principal control parameters for the generator speed. An appropriate system characteristic is shown Fig. 4. Equilibrium is established when the operating point on the characteristic matches the operating point on turbine power/speed curves. For any given shaft speed, there exists a corresponding optimal power. One such optimal speed/power relationship, is shown dashed in Fig. 4, could be stored in the memory of the controller and compared with actual operating points to decide the amount of control variable adjustment required.

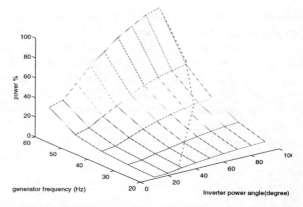

Figure 4: Characteristics of generator-power electronics conversion system

Reactive power control

The reactive power is generated by the circulation of current through the inverter switches. The inverter is the source and control of the reactive power, the front-end AC/DC rectifier carries the real power only. The reactive power can be controlled to meet the system requirements such as a desirable power factor or a required voltage level at the ac terminal of the inverter.

Controller

A suitable controller block diagram is shown in Fig. 5. The inputs are machine speed, dc link voltage and current, ac network voltage and reactive power requirement. The outputs are the power angle (δ) and the on/off ratio of switch S_D (N_k). It should be noted that wind velocity is not required as an input signal for the control system.

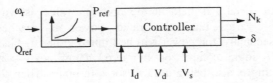

Figure 5: Schematic power controller

HARMONIC POLLUTION

Harmonic reduction

The system described can provide efficient power conversion and fast control. However, semiconductor switching causes harmonic pollution. The VSI is a source of voltage harmonics at the inverter ac terminal. The resulting current harmonics injected into the grid depend upon the converter configurations, switching patterns, system parameter and operating conditions. Current harmonics in the grid are becoming a severe nuisance and are justly regarded as serious pollution.

A quasi-square, six-pulse inverter has a simple circuit configuration and leads to low switching loss but causes high levels of harmonic pollution.

Harmonic distortion can be reduced by using higher switching frequency and patterns such as selective harmonic elimination (SHE) or sinewave PWM. SPWM reduces low order harmonics and SHE eliminates selected low order harmonics. Higher order voltage harmonics, are increased but these create little pollution in the inductive dominated power system and are easily filtered. The higher frequency switching techniques could be directly implemented on the basic three-phase bridge circuit but would lead to higher switching power losses and a compromise between efficiency and pollution is required.

Multiple inverter systems are a widely-used alternative. Separate transformers or multiple secondaries combine the outputs from a number of bridges, each rated at a fraction of the system total power. The voltages generated by individual inverters are equal in magnitude and harmonic content, but when combined in the phase shifting transformer system, many of the harmonics are cancelled. If the bridges are perfectly balanced and the transformer effective turns ratios are exactly equal then only harmonics of pk ± 1 order are presented to the grid, where p is the total pulse number and k is a positive integer (1,2,3,......).

The total current harmonic distortion (TCHD) is the normal harmonic pollution index and is defined by

$$TCHD = \frac{I_{dis}}{I_{(1)}} \times 100\% \qquad (2)$$

where $I_{(1)}$ and I_{dis} are the rms value of the fundamental component and harmonic components of the current.

Simulated line-line voltage waveforms and harmonic spectra for 6-pulse quasi-square, 6-pulse SHE switching

(5th, 7th, 11th and 13th harmonic elimination) and 24-pulse inverter systems are shown in Fig.6. The TCHD characteristics of these systems are presented in Fig. 7. Obviously, as the fundamental current tends to zero, the TCHD becomes infinite.

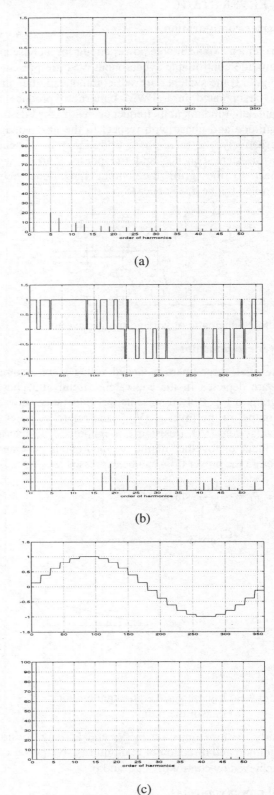

(a)

(b)

(c)

Figure 6: Inverter voltage and harmonic spectra
(a) Six-pulse (b) SHE switching (c) 24-pulse inverter

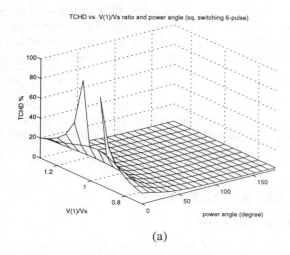

(a)

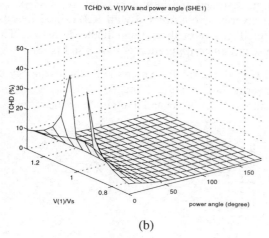

(b)

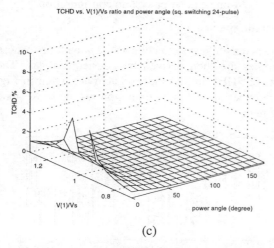

(c)

Figure 7: TCHD vs. power angle and $V_{I(1)}/V_s$ ratio
(a) Six-pulse (b) SHE switching (c) 24-pulse inverter

It can be seen that the 24-pulse inverter system gives excelent TCHD performance. Using selective harmonic elimination in a multi-pulse system has been studied and no significant benefit found in respect of TCHD performance.

Semiconductor power losses of an IGBT inverter

During a conduction period, the energy dissipated in each switch (IGBT or diode) can be written as:

$$E = \int_0^t v_{ce}(i)i(t)dt \qquad (3)$$

where v_{ce} is the voltage across the semiconductor, i is the current carried. The mean power dissipated may be obtained by multiplying energy by frequency.

Equation 3 encompasses both conduction and switching (turn on, turn off, and recovery) losses. These components may be calculated separately taking account of the current waveforms such as Fig. 8.

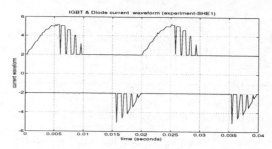

Figure 8: Current waveforms of semiconductors

(upper-IGBT, lower-diode)

Conduction energy losses, The instantaneous power loss of the IGBT and diode is the product of the current through the semiconductor and the on state voltage drop which is a function of the current. The conduction energy losses can be calculated by summing the integrations of the instantaneous power loss in each conduction period.

Switching energy losses, The turn on, turn off and reverse recovery energy loss of the semiconductors can be calculated by using the current waveform with the energy loss curve on the data sheet or with simplified formulae which give the reasonable accuracy[7].

The simulated current waveform and corresponding power loss of an IGBT 3-phase bridge inverter is presented below along with the harmonic performance. The conduction losses of a SHE six-pulse system are nearly the same. The current waveform of the basic 6-pulse system is rich in harmonics which results in higher (about 6%) conduction losses. The switching loss is proportional to the switching frequency. For the IGBT inverter, switching losses exceed conduction loss for switching frequencies greater than about 6kHz.

Comparison of harmonic reduction methods

The schemes listed in Table 1 have been investigated by computer simulation and are compared in Tables 2 and 3. For calculation of TCHD it is assumed that:

• The coupling reactance (inverter to grid) is 0.1 pu

- The inverter generates 1 pu reactive power, and zero real power (the worse case for TCHD).

Table 1 Inverter harmonic reduction options

Identifier	Particulars
Square wave switching	
SS6	6- pulse inverter system
SS12	12- pulse inverter system
SS24	24-four pulse inverter system
Selective harmonic elimination	
SHE1	eliminate 5th, 7th, 11th, and 13th harmonics.
SHE2	eliminate 5th, 7th, 11th, 13th, 17th and 19th harmonics.
SHE3	eliminate 5th, 7th, 11th, 13th, 17th, 19th, 23rd and 25th harmonics.

Table 2 Comparison of power losses

Identifier	switching power loss (W)	conduction power loss (W)	switching frequency (Hz)
SS6	5.7	872.8	50
SS12	5.7	561.2	50
SS24	5.7	405.5	50
SHE1	47.1	829.6	450
SHE2	79.4	822.1	650
SHE3	102.7	822.1	850

Table 3 Comparison of harmonic performance

Identifier	TVHD %	TCHD % X_s=0.1 pu	Number of transitions /device/cycle
SS6	30.55	51.02	2
SS12	14.95	11.61	2
SS24	7.434	2.83	2
SHE1	51.14	24.47	18
SHE2	51.22	18.88	26
SHE3	50.69	15.29	34

The loss calculations presented are for typical conditions: V_{DC} = 600V, $I_{AC(1),RMS}$ = 100A, This leads to the multiple-pulse inverters being under loaded.

Alternatively, if the bridges are all fully loaded then the rating of say the 24-pulse system SS24 is four times that of SS6 and its efficiency is the same. The device characteristics given in reference 7 have been used.

It can be seen that the multiple pulse inverter is more effective at reducing the TCHD. In the 24-pulse case, an external filter may not be required since the TCHD should meet statutory requirements.

EXPERIMENTAL STUDIES

Experimental system

A laboratory model of the proposed conversion system has been designed and built. The synchronous voltage source inverter is constructed with IGBTs and may be configured in any of the arrangements discussed. The 24-pulse circuit configuration is shown in Fig. 8.

The designed control system can keep the frequency of the inverter output voltage locked with the grid and regulates both the phase and magnitude of the output.

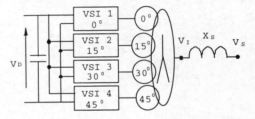

Figure 8: 24-pulse inverter system configuration

Harmonic spectra

A data acquisition system collects the waveforms whose harmonics are then derived by Fourier analysis. The results of SHE1 and 24 pulse system are given in Fig. 9. The experimental and simulation results show good agreement.

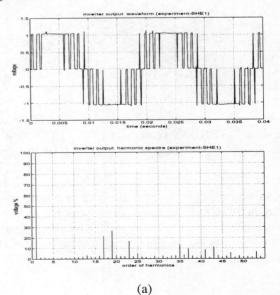

(a)

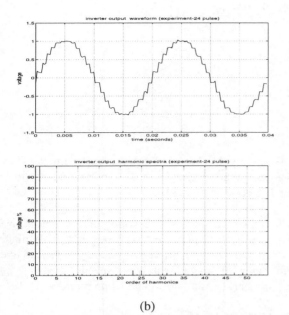

(b)

Figure 9: Voltage&harmonic spectra (experimental)
(a) SHE switching (b) 24-pulse inverter

DISCUSSION

In the variable speed wind energy conversion system, the generator and grid are de-coupled by the power electronics converter. The interface performance, such as voltage regulation, reactive power control and power quality mainly depend on the inverter. This gives more freedom to the modular PM generator designer bringing cost and performance benefits. It also provides economies by removing the need for special synchronising and damping equipment of the direct-drive PM generator.

The system may operate at any power angle without losing synchronism. The reactive power generation is de-coupled from the generator, which allow the capacity of the generator-rectifier to be fully utilised in the real power conversion. The system can provide fast dynamic control of reactive power to compensate for variable power factor loads and adjust system voltage.

In the presented cases, the main power losses are conduction power losses which is approximately proportional to the square of the load current. In the multi-pulse inverter system, each inverter carries only a fraction of the total load current, consequently, the semiconductor losses are small.

The drawback of the multiple pulse inverter system is the multiplication of devices and their associated circuits. However, in a windfarm, the effect of a multiple-pulse inverter system can be achieved at minimal cost by phase shifting simple six pulse inverters connected to the individual wind turbines.

Alternatively the multi-pulse scheme allows the desired rating to be assembled from several smaller units which may be advantageous for the development of the very high power wind turbines currently being discussed.

CONCLUSIONS

The paper has discussed the application of the voltage stiff power electronics converter to interface a variable speed PM generator to the grid. The proposed voltage source inverter and control scheme enables the system to track the maximum power coefficient of the turbine as wind speed varies and hence to achieve optimal energy capture from the wind.

The reactive power generated by the inverter can be controlled to keep the inverter output at a desired power factor or a voltage level.

High-quality power is delivered to the system by multiple-pulse inverter operation or by a multiple inverter configuration in a windfarm.

ACKNOWLEDGEMENTS

The first author wishes to thank the School of Engineering, University of Durham for providing the facilities and financial resources.

REFERENCES

1. Halliday, J. A.: "Wind energy: an option for the UK?", IEE Proc. A, 140, 1, Jan 1993, pp 53-62.

2. Ralls, K. J. "The growth of power electronics in electrical power transmission systems", Power Eng. Journal, February, 1995, pp 15-23.

3. Spooner E., Williamson A.C., Thomson L.: "Direct-Drive, Grid-connected, Modular Permanent-Magnet Generators", British Wind Energy Assoc. Conference, Stirling, June, 1994

4. Spooner, E., Williamson, A. C.: "Direct-coupled, permanent-magnet generators for wind turbine applications", Submitted to IEE Proc-B Electric Power Applications.

5. Spooner, E., Williamson, A. C.: "Modular design of permanent-magnet generators for wind turbines", Submitted to IEE Proc-B Electric Power Applications.

6. Chen, Z., Spooner E.: "A Modular, Permanent-Magnet Generator for Variable speed Wind Turbines", IEE International conference EMD'95, Durham, September 1995.

7. Casanellas, F.: "Losses in PWM inverters using IGBTs", IEE Proc- Electric Power Applications, vol. 141, No. 5, September 1994.

INTEGRATION OF WIND TURBINES ON WEAK RURAL NETWORKS

L M Craig*, M Davidson**, N Jenkins*, A. Vaudin**

*UMIST, UK
**National Wind Power Ltd, UK

INTRODUCTION

Under the third round of the non-fossil fuel obligation (NFFO) in March 1995, the UK government agreed contracts for nearly fifty new wind generation projects, with prices as low as 4p/kWh. When compared with prices of 11p/kWh under the second round of NFFO contracts in 1991, it is clear that wind energy is rapidly becoming a commercial source of electricity. Lower prices for wind generated electricity tend to concentrate wind farm developments at high wind speed sites, which, in the UK, are usually upland areas with low population density, remote from a strong electrical connection point. The high capital costs of reinforcing the network, together with the difficulty of obtaining planning permission for new overhead lines, encourage maximum use of the existing network infrastructure. Rural electricity networks in the UK are characterised by long medium voltage lines to distributed loads, which are predominantly single phase. This leads to a low fault level and low X/R ratio at the point of connection. These conditions, combined with the high wind turbulence intensity which is often associated with upland terrain, provide the least favourable circumstances for the quality of output power from wind turbines.

Two case studies have been conducted to investigate the power quality of wind turbines connected to weak rural networks. The first was at a single wind turbine connected to the 11kV network in County Antrim, Northern Ireland. The second study was at a wind farm of 24 wind turbines, connected to the 33kV distribution network at Cemmaes in mid-Wales. The impact of the wind turbines on the network and the effect of network disturbances on wind turbine operation were evaluated in terms of steady state voltage levels, dynamic voltage variations, flicker and voltage unbalance.

SINGLE WIND TURBINE

An intensive programme of measurements was carried out at a single 300kW wind turbine, connected to the Northern Ireland Electricity 11kV network via a 415/11000V transformer. The site is supplied by 11kV overhead lines from the nearest substation, over a distance of 8 km. The 11kV fault level at the site is usually 12MVA, with an X/R ratio of 0.6. There are two alternative feeders from the 33/11kV substation, which give 11kV fault levels of 8MVA and 15MVA

respectively at the wind turbine transformer. These feeders were switched to enable measurements to be made at a range of site fault levels.

The power quality of the wind turbine output was assessed by measuring instantaneous and rms voltage, rms current, power, reactive power and wind speed over the range of operational wind speeds, during wind turbine start up, shut down and while running [1].

Steady State Voltage

Measurements of the rms voltage at the 415V busbar, with the wind turbine out of operation, showed that the site voltage varied on a daily cycle by over 4% between maximum and minimum 10 minute average values. These steady state voltage variations can be accounted for by changes in load on the network. During a typical 24 hour period, the lowest voltage occurred around 8am, corresponding to a peak load demand. Voltage levels were highest between midnight and 6.30am when load was very light. Due to the very low X/R ratio, which results from the relatively high resistance of the 11kV overhead lines, changes in both active and reactive power flows have significant effects on the steady state voltage changes. The steady state voltage changes caused by the wind turbine due to different levels of power output were observed, superimposed on this daily variation. The percentage steady state voltage changes over the range of wind turbine power output are shown in Figure 1. The figure shows the percentage voltage changes at the 415V busbar over the range of wind turbine output, for the 8MVA fault level, together with the voltage variations calculated by a fast decoupled load flow algorithm. There is a very close agreement, with a maximum difference of 0.3%.

Dynamic Voltage Fluctuations

Dynamic voltage variations are caused by wind turbine starts, shutdowns, and by fluctuating power output resulting from wind gusts and the effects of wind shear and tower shadow on the rotating aerodynamic rotor.

At start up, the wind turbine is connected to the network by means of an anti-parallel thyristor softstart, which limits current inrush to approximately rated current. Measurements at the wind turbine showed that the percentage voltage drop at the 415V terminals varied between 1.1% and 1.5%, which corresponds to a voltage drop of less than 0.5% at the site 11kV busbar. According to Engineering Recommendation P28 [2], the number of turbine starts would be unrestricted, for a disturbance of this magnitude.

At shutdown, the induction generator remains connected to the network until the aerodynamic rotor comes to a standstill. A one way clutch in the drive train prevents the generator from driving the aerodynamic rotor. The voltage therefore drops as the induction machine draws a small motoring current for a few seconds, and then rises to a new steady state value when the machine is disconnected. These changes are illustrated in Figure 2. The corresponding voltage drop at the 11kV busbar was approximately 0.5%.

The effect of tower shadow and wind shear on the wind turbine output power can be seen in Figure 2, where there are obvious power fluctuations, prior to shutdown, at the frequency at which the blades pass the tower (blade passing frequency, or 2p for the two bladed machine). Although the frequency at which this occurs, 1.6Hz for this fixed speed wind turbine, is below the frequency range to which the human eye is most sensitive, the operation of the wind turbine was observed on the flickermeter connected at the wind turbine terminals.

Flicker was recorded at the wind turbine 415V busbar, and at the 11kV busbar at the 33/11kV substation. Background flicker levels were very high, due to a significant 4Hz frequency component, which was present in the voltage traces during the operating hours of a local quarry. Outside the quarry's working hours, the flicker due to wind turbine operation varied from 0 to a P_{st} of 0.25. Flicker levels at the 33/11kV substation were unaffected by normal wind turbine operation. However, calculations predict that at the wind turbine 11kV busbar, the corresponding flicker would exceed a P_{st} of 1 at the maximum operating wind speed of the wind turbine [3].

Unbalance

The background voltage unbalance at the wind turbine site varied between 0.5% and 1.5%, which is not unusual for a rural network with a high percentage of single phase loads. Operation of the wind turbine was observed to reduce the network voltage unbalance by approximately 0.2%. The negative phase sequence currents which are induced in the generator rotor can, however, lead to significant heating when the network voltage unbalance is continuously high, and over-temperature trips occurred when the generator was first commissioned. The generator was designed for a voltage unbalance of 2%, and output derating is necessary if unbalance exceeds this level.

CEMMAES WIND FARM

Cemmaes wind farm consists of 24 300kW wind turbines, connected to Manweb's distribution network at 33kV. The 33kV fault level is 78MVA, with an X/R ratio of 1.8. The wind turbines are of similar design to the single wind turbine considered above. Each wind turbine generates at 660V and is connected to a 660/11000V transformer at the base of the tower. Site power is collected by 11kV underground cables and is supplied to the distribution network via a 33/11kV transformer at the wind farm substation.

Data from the wind farm was collected from the site SCADA system, which provided 10 minute summary data on wind turbine and wind farm output, as well as wind turbine operational status. Rms power, reactive power, voltage and current were measured by Data Acquisition Systems (DASs) located at two wind turbines and at the substation. A flickermeter, monitoring 33kV flicker levels, and a disturbance recorder were installed at the substation

Steady State Voltage

The 33kV rms voltage was measured over an extended period when the wind farm was out of operation. The 10 minute average of the rms voltage at the 33kV busbar varied between +3% and -10% of nominal voltage. Figure 3 shows the variations in 33kV voltage corresponding to the wind farm output power. The background voltage variations distort the relationship between output power and voltage, but the trend of increasing 33kV voltage with increasing output is clearly seen. Voltage changes of -6% and +3% between minimum and maximum generation were recorded, which are less than the magnitude of the voltage changes with the wind farm out of operation.

Dynamic Voltage Fluctuations

The effect of wind turbine starts on the busbar voltage can be seen in Figure 4, which shows the wind farm reactive power demand and the 33kV busbar voltage following a reconnection of the wind farm after a G59 protection trip. The connection of the first wind turbine at 107s gives a voltage dip of 1% at the 33kV busbar. The connection of subsequent generators gives a smaller disturbance, of approximately 0.5%.

The variation in reactive power at the blade passing frequency (1.6Hz), which can be observed in Figure 4, correspond to the power fluctuations at this frequency. The magnitude of these fluctuations varies between less than 1% and 10% of the reactive power demand. Observations at this and other wind farms [4] indicate that wind turbines can fall into synchronism in certain conditions, which amplifies the disturbance at the blade passing frequency. Generally, it is assumed that individual wind turbine power fluctuations are uncorrelated and, in a wind farm, a smoothing of output power variations will occur. Investigations of the phenomenon of synchronous operation are still proceeding. Certain types of wind turbine are more susceptible than others to these power fluctuations. Variable speed wind turbines with fully controlled converters between the network and the generator can reduce cyclic power variations to an insignificant level.

Despite the obvious power fluctuations, the flicker measured at the 33kV busbar was observed to decrease as the wind farm output increased. Figure 5 shows the mean flicker level (P_{st}) plotted against wind farm output. The average P_{st} drops from a maximum of 0.2 at minimum wind farm output to a minimum of 0.15.

It can be shown that voltage flicker generated by a wind farm is very sensitive to the network X/R ratio at the connection point [3]. A minimum in flicker occurs when the X/R ratio is approximately 1.7, which is close to the ratio at Cemmaes. The combination of a favourable X/R ratio and the increased fault level when the wind turbines are connected, are probable reasons for the reduction in flicker when the wind farm is operating.

Unbalance

Figure 6 shows the average of the 10 minute mean negative phase sequence voltages, plotted against wind farm power output. The negative phase sequence voltage decreases from a mean of 1% at zero generation to a minimum of 0.2% as the wind farm output increases. The heating effect of the negative phase sequence currents has not affected wind farm performance to the extent that derating of the generators is required.

CONCLUSIONS

Steady state voltage levels rise with increasing power output, both for the single wind turbine connected to the 11kV network, and for the wind farm on the 33kV network. The magnitude of the voltage rise is less than the background variation without the wind turbines connected.

Dynamic voltage fluctuations due to wind turbine start up and shutdown are within the limits recommended by utility guidelines. However, in very high winds, flicker from a single wind turbine connected to a weak network may exceed recommended limits. Dynamic voltage fluctuations at the wind farm were observed to be less significant than for the single wind turbine, due to the smoothing effect of the superposition of uncorrelated fluctuations, although occasional occurrences of synchronous rotor operation have been observed.

Network voltage unbalance is reduced by the induction generators used in fixed speed wind turbines. When network voltage unbalance is very high, generator output derating may be required, due to the heating effect of the negative phase sequence currents. This has not been necessary at the wind farm, where the background voltage unbalance is approximately 1%.

Experience of operating wind farms in the UK and Europe to date has shown that embedded wind generation will cause minimal disturbance to distribution system operation. As the level of network penetration increases, greater flexibility in system planning will be required to reap maximum benefit from the potential provided by local generation to meet local load demand.

ACKNOWLEDGEMENTS

The work reported in this study formed part of projects funded by the DTI, and managed by the Energy Technology Support Unit.

REFERENCES

1. Craig LM and Davidson M, 1995
   Integration of Wind Turbines onto the Electrical Network
   DTI Energy Technology Support Unit, Report W/33/00265/REP

2. Engineering Recommendation P28
   Planning limits for voltage fluctuations caused by industrial, commercial and domestic equipment in the UK
   The Electricity Association, 30 Millbank, London SW1P 4RD

3. Gardner P, 1993
   Flicker from Wind Farms
   Proceedings of BWEA/RAL workshop on Wind Energy Penetration into Weak Electricity Networks

4. Stampa A and Santjer F, 1995
   Synchronism of grid connected wind energy converters in a wind farm
   DEWI Magazin No. 7 (in German)

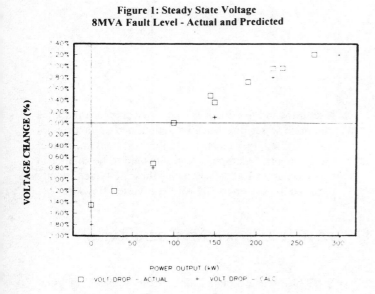

Figure 1: Steady State Voltage
8MVA Fault Level - Actual and Predicted

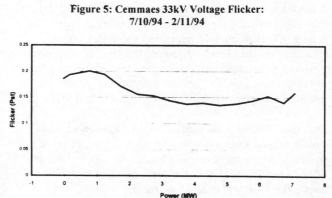

Figure 4: Reactive Power and Voltage
Following Wind Farm Reconnection

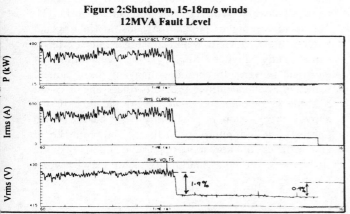

Figure 2:Shutdown, 15-18m/s winds
12MVA Fault Level

Figure 5: Cemmaes 33kV Voltage Flicker:
7/10/94 - 2/11/94

Figure 3: Mean 33kV Voltage and Wind Farm Power:
7/6/95-22/6/95

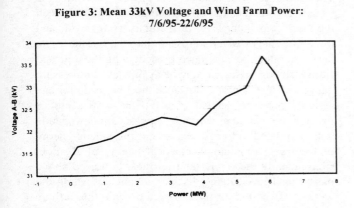

Figure 6:Cemmaes 11kV Voltage Unbalance:
7/10/94 - 2/11/94

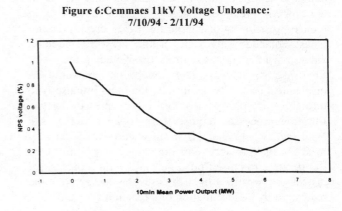

EMBEDDED MINI-HYDRO GENERATION IN THE WATER SUPPLY INDUSTRY

A R Wallace

University of Edinburgh

There is widespread potential for generating modest amounts of hydro power at treatment plants in the United Kingdom water supply industry. This paper identifies some of the potential sources of mini-hydro power, reviews the applicable mechanical, electrical and control equipment, and considers some of the factors influencing technical and economic viability of power generation at these sites. The institutional regulations applicable to new plants are identified.

OPPORTUNITY FOR POWER GENERATION

The demand for water from treatment plants can exceed 20 million gallons/day (1.05 m³sec⁻¹) and in many cases the water is delivered to the site by gravity, containing a significant amount of energy. The delivery head can vary from a few metres to over 100 metres. It is frequently necessary to reduce that water pressure to the level required by the downstream treatment processes, and this creates the opportunity for micro-or mini-hydro power generation where the installed capacity could be between 10 and 1000kW.

Different filtration processes require higher or lower water pressures to pass water through them. Revolving strainers and drum-screens followed by slow sand-bed filters operate with the delivery head a few metres above the free water level of the filter beds. Active filtration by anaerobic blankets requires a similar delivery head. It is necessary in these cases to break all pressure in the water delivered, and re-establish delivery head by discharge into elevated tanks at atmospheric pressure. Largely, present practice is to reduce pressure by discharge through break-pressure tanks, colander valves or in-line pressure reducing valves. The energy is entirely dissipated in the water flow itself, but may alternatively be recovered as mechanical power by the installation of an impulse-type turbine. Rapid-gravity or other pressure-driven filtration processes require there to be a higher site head and, where there is a sufficiently high delivery head, in-line pressure reducing valves may be installed to operate into closed pipes or tanks with a prescribed discharge pressure. In such cases the energy could be recovered by the installation of a reaction-type turbine.

There are a number of other opportunities for developing mini-hydro power associated with the conveyance of water between intermediate storage reservoirs and treatment works. At compensation sites minimum water flow must be maintained in the original watercourse to meet wildlife or environmental requirements. Compensation sets tend to be smaller in capacity, up to 100kW say, since they are typically only required to pass up to 10% of the diverted flow.

POTENTIAL FOR POWER GENERATION

In 1989 it was estimated that, of the 36.3MW of mini-hydro potential in England and Wales, 16.8MW was available for development at 30 sites within the Water Industry. As part of the same study it was estimated that 2MW out of the 238MW potential in Scotland could be developed at 12 water treatment works [1]. The relatively higher population densities in England require more treated water at larger sites with a consequently higher capacity for generation of electrical power. In addition, many of the treatment works in Scotland utilise pressure filters and it is not possible to recover the power. At the time of the above survey economic viability was based on an internal rate of return of 10%. The commercial circumstances are now more opportune and many of the sites then identified as marginally viable may be worth reconsidering.

SELECTION OF GENERATING PLANT

The gross head available for power generation is usually recorded as the difference in elevation between the mean free water level of the delivery reservoir and the centre line of water admission into the turbine. Net head can either be calculated by deducting the pipe and fitting friction losses, or may be available from pressure gauge measurements. Allowance must be made for any residual pressure needed to drive the water into or through the treatment works. Impulse turbines operate with ventilated tailraces, converting the kinetic energy of the free jet(s) into mechanical energy of rotation. Consequently the runner and jet must be raised above the on-load level of the discharge channel feeding the works. Reaction turbines operate with the turbine casing flooded and the pressure difference across the runner develops output torque. Inlet and shaft centre lines are usually set slightly lower to keep the turbine draft tube under positive pressure. The flow rates

Opportunities and Advances in International Power Generation, 18–20th March 1996, Conference Publication No. 419, © IEE, 1996

against which the turbine must be selected are usually known and this avoids the uncertainty normally associated with hydrological estimation of flow. The output power of the turbine may be estimated as

$$P_{out} = \rho.g.h.Q.\eta \quad \text{Watts}$$

where ρ is the density of water (1000 kg.m^{-3}), g is the acceleration due to gravity (9.81 m.sec^{-1}), h is the net head (m), Q is the water throughput (m^3.sec^{-1}) and η is the efficiency of the turbine which might initially be assumed as 0.85. The turbine must be selected, subject to the constraints on setting and residual pressure, to develop maximum power at the highest possible synchronous speed for generation N (rev.min^{-1}). The power developed, operating head, and say four pole shaft speed, 1500 rev.min^{-1} are used to estimate a *site specific speed* N_s from

$$N_s = N.\frac{P_{out}^{0.5}}{h^{1.25}}$$

The site specific speed may be compared with the ranges in specific speed available from the different types of turbine tabulated below.

Turbine Type	Specific Speed N_s
Single jet Pelton	10 - 28
Twin jet Pelton	15 - 40
Single jet turgo	48 - 72
Twin jet turgo	68 - 101
Francis	70 - 300
Axial flow	350 - 700

Manufacturers tend to propose turbines based on a selection of well-proven runners of distinct specific speeds of whose efficiency they are confident. A particular machine may be applied over a small range in head and flow with acceptable variation of speed and efficiency. Reaction turbines offer better peak efficiency than impulse machines and are suited to applications where the site demand for water is unlikely to vary, but above or below design flow there is a loss of efficiency. An advantage of single- or multi-jet impulse machines is that the water is admitted into the runner through one or more identical nozzles which can provide full- and part-flow operation. Controlling flow by spears in the nozzles maintains the quality of each jet and impulse turbines offer better part-flow efficiency at sites where the demand for water may vary.

In cases where the net head is low, even high specific speed machines may require to be connected to four or six pole generators by speed-increasing gearboxes. Generators may be chosen from self-excited synchronous type or singly-excited induction machines deriving excitation from the grid. Brushless

synchronous generators are relatively more expensive than induction machines since they include excitation sources, voltage regulation systems, ac exciters, rotating rectifiers and salient-pole rotors with bar- or wire-wound main field coils. They are, however, able to self excite and operate in the absence of the grid. Once synchronised they may be operated overexcited at variable lagging power factors to export active (kW) and reactive (kVAr) power to the grid. In the presence of a strong grid induction motors may be selected to operate at super-synchronous speeds and export active power to the grid. They must be magnetised from the grid and will operate at a leading power factor which varies with the power produced by the turbine. Power factor correction systems for induction motors used to be designed to avoid self-excitation of the machine, but self-excitation of stand-alone induction machines is now fairly common using microprocessor controlled power electronic regulators and capacitors.

It is a feature of all turbines that loss of load can result in overspeed, defined by the turbine type. It is imperative the generator rotor body, field winding, all rotating components, and bearings are designed and constructed to operate for sustained periods at full turbine overspeed. In addition the shaft system and bearings must be capable of absorbing continuously the axial and radial thrusts produced by the turbine. These are fundamental requirements for safe and reliable operation, and must be incorporated in specifications.

CONTROL EQUIPMENT

At some sites featuring compensation sets in the 10-50 kW range, Pelton and turgo-impulse machines operate with shaft mounted centrifugal governors acting on jet deflectors. Larger or more modern machines are now provided with mechanical-hydraulic governors or electro-hydraulic governors, which act on the jet deflector in the case of impulse turbines or the guide vanes in the case of Francis machines. These governors include manual or motorised adjustment of speed setting, and adjustable speed-droop, which enables automatic synchronising, and predetermined load share with the grid or other machines. In recent years electronic load controllers have found application at smaller sites where the hydro plant is not synchronised to the grid.

At many of the early water treatment works, where site load, or water throughput did not vary greatly, and operating staff were available for start-up or shut-down sets were provided with manual control only. Nowadays generating plant is usually designed for unmanned operation with fully automated start and stop control under control of relay logic or programmable logic controllers. For fully automatic control, valve actuators must be motorised, speed and voltage presets must be motorised and the control system must include automatic and check synchronising equipment.

The plant must be monitored while energised and protected against mechanical or electrical failure. Mechanical failures such as increased generator stator or bearing temperatures, or increased vibration levels may be used to unload the plant and initiate a pre-determined sequence shutdown. More serious electrical failures must be detected and used to initiate rapid disconnection of the generator from the grid. Sequence control must thereafter shut down the turbine and auxiliary plant.

Following such a trip, it is possible with some impulse turbines to lock the deflector in the jet and pass the required flow through the turbine with the runner stationary. With reaction turbines this is not possible and in most cases a pressure-reducing valve is used to bypass the turbine and maintain head for the restart. Where the turbine is supplied from a relatively low volume intermediate storage reservoir or balancing tank there is the possibility that mis-match between the inflow to the tank and the water drawn off to pass through the turbine or pressure reducing valve will cause draw down of head or spillage. Microprocessor based level controllers can be specified and made to act on the turbine spear (or bypass) valve to maintain head within prescribed limits.

TECHNICAL CONSTRAINTS

In older water supply systems the maximum water pressure in the supply network may be limited, and in many rural supply systems feeding large cities the supply penstocks can be up to 30km long. Pressure rise induced as the flow of water is interrupted has to be strictly limited and the closure (and opening) times of in-line valves of up to ten minutes are common. Impulse machines are often installed at the end of such pipelines since they respond to loss of load by deflecting water into the tailrace without immediate closure of valves. Speed-rise is avoided but with a reaction turbine this will only be limited by the governor closing the guide vanes in the spiral case in seconds rather than minutes, resulting in substantial pressure rise. Reducing the rate of closure of the guide vanes allows the speed of the turbine and generator to rise and it is always necessary to trade off pressure and speed rise in the specification of reaction turbines. Raising the moment of inertia of the generator is the traditional way of limiting speed rise. With proprietary generators in the mini-hydro range it is not normally possible to specify additional inherent inertia, and provision of a flywheel increases the complexity and costs of bearing and mounting arrangements.

For smaller capacity medium-voltage installations the full load current produced by the generator may be relatively low. The three phase symmetrical short circuit current from the generator may represent a fault level of only a few MVA, but the plant must be designed for safe operation and clearance of the total fault level at the point of connection. This may vary from a few MVA in medium voltage rural situations to several hundred MVA in more heavily reinforced high voltage areas. To provide adequately fault-rated switchgear it may be necessary to connect to the grid at high voltage. Step-up transformers and high voltage switchgear may become necessary to export what could be a relatively small amount of power to the grid.

Local supply authorities may impose their own requirements in terms of system protection, operation and earthing, but in many cases they base their minimum requirements on Electricity Association Recommendation G59/1, [2]. It is a requirement of G59/1 that the generator be provided with dedicated protection against: loss of grid; undervoltage; overvoltage; underfrequency; overfrequency. In addition it is essential that the plant be protected against: overcurrent and earth fault; reverse power; and either standby earth fault or neutral voltage displacement. This is prescribed to protect the grid against local disturbance, the plant from internal or grid-located electrical failure and avoid the connection of an incoming generator to an unhealthy grid. At smaller plants the protection can become a significant part of the overall cost of control and switchgear.

Generator neutral earthing must recognise the nature of the earthing of the supply authority's low or high voltage system to which the generator is connected. In addition, connection of local earths and equipotential bonding to the supply authority earth is subject to their agreement. It may be necessary to provide a local earth which fulfils the requirements of the 16th Edition of the IEE Wiring Regulations and BS7430. Sites in treatment works available for the construction of power houses are seldom in virgin ground and provision of a satisfactory earth should be considered at the outset.

INSTITUTIONAL FRAMEWORK

Any new installation, even on an existing site, will require express planning permission and the application will be judged in terms of the normal requirements for new constructions. Additionally, environmental planning measures are likely to be applied including conservation of wildlife, local amenity, and rights of way. Assessments of the revenue from the plant should include any charges becoming due under the recovery of land rates. It may be necessary to obtain a water abstraction licence under Part II of the Water

Resources Act 1991, together with permission to discharge back into the watercourse, but installations confined to within a treatment works may avoid this. Financial assessments should include any abstraction charges becoming due.

As a consequence of the England and Wales (and Northern Ireland) Non Fossil Fuel Obligations (NFFO 1-3) and the Scottish Renewables Obligation (SRO1-2), the Public Electricity Licensees are compelled to purchase electricity generated by a further 400MW capacity of renewable energy plant. The increased tariff available (around £40/MWh) is intended to encourage investment in economically marginal mini-mini-hydro power projects. A Power Purchase Agreement with the REC is a pre-requisite in applying for support under the NFFO or SRO. In the mini-hydro range it may be necessary to obtain confirmation from the Secretary of State for Energy that the plant falls within The Electricity (Class Exemptions from the requirement for a licence) Order 1990, and that a licence is not required [3].

ECONOMIC EVALUATION

The most straightforward method of computing the economic virtue of investment in mini-hydro plant is to consider the payback period in years that the net revenue from the plant (yearly income from energy sales less all annual operating expenses) will take to recover the capital invested. This period is sensitive to inflation rates, interest rates on borrowed capital, and tariffs applied. Over the trading period the revenue and expenditure may be increased to take account of inflation, but in later years the actual sums of money may become difficult to relate to the present value - on which basis the investment capital may be issued. For this reason, present-worth (PW), net present value (NPV), or discounted cash-flow (DCF) methods are often employed.

Irrespective of the means of analysis it is fundamentally important that the capital costs (and consequent interest charges) are minimised and the revenue from the energy produced is maximised. An advantage of developments at existing structures or at water treatment works is that the intake, penstock and discharge pipework may already exist and the scheme capital costs are reduced. To maximise the revenue from energy sales the plant should be operated at maximum capacity for the longest possible periods of time, commanding the highest tariff for energy produced and sold to the REC. Increased tariffs available under the NFFO assist project viability greatly, but without this exported active power from synchronous generators to the RECs attracts a buy-in rate typically less than £17/MWh. Where it is possible

to reduce energy purchase from the Regional Electricity Companies the offset value of generated power or avoided costs of purchase can vary up to £70/MWh. It therefore assists the economic viability of projects if there is a high site demand for energy. This is usually the case at water treatment works where there may be pumps, compressors, sludge presses, heaters and high lighting load. Currently no allowance is made for the provision of reactive power from synchronous generators, but some measure of benefit may be obtained by improving the site power factor to unity and avoiding reactive power supply charges.

CONCLUSIONS

There is significant potential for developing mini-hydro power within the water industry, and depending on the increasing demand for treated water into the next century, the opportunity for installing economically viable plants is set to increase.

The technology which may be installed at these sites is mature, reliable and readily available within the United Kingdom. Many of the institutional barriers to development at other sites do not apply at sites owned and operated within the water industry. The viability of investment in these plants depends on the availability of site electricity demand and a purchase agreement with the REC under the terms of the NFFO.

Depending on the extent of new works many schemes can be demonstrated to be economically viable, recovering their investment in acceptably short periods of time, and thereafter generating revenue from the recovery of energy which otherwise would be dissipated.

REFERENCES

1. SCEL, 1989, "Small Scale Hydroelectric Generating Potential in the UK", UK Department of Trade and Industry ETSU, report number ETSU-SSH-4063.

2. The Electricity Association, 1991, "Recommendations for the Connection of Embedded Generating Plant to the Regional Electricity Companies' Distribution Systems", Engineering Recommendation G59/1

3. Brandler, A, 1993, "UK Department of Trade and Industry ETSU, Project Financing Renewable Energy Schemes", ETSU-K/FR/00028/REP

AN ADVANCED ELG FOR CONTROL OF SMALL SCALE HYDROELECTRIC GENERATION

D S Henderson

Napier University, UK.

INTRODUCTION

Micro hydroelectric sets are defined as those with unit ratings of less than 100 kW. They are often situated in remote communities, particularly in developing countries. As they are often isolated from grid networks their technical characteristics are such that they require a governor to maintain the frequency at an acceptable level for the users. The specification for a rural electricity supply as defined by Woodward (1) is a lot less rigorous, or rigid, as that for Western countries. The communities which install these sets usually have limited finance and limited skilled labour, if any, to operate and maintain the equipment. Unfortunately, as the rating of hydroelectric plant decreases then the cost per kW increases, Wallace (2). Hence, for a community to afford a micro hydroelectric generating set, the capital cost of the plant must be as low as possible and the plant must be as simple to install, operate and maintain as is possible.

A key item of the plant is the governor. Traditionally, speed governors such as the mechanical-hydraulic type have been installed which adjust the speed of the set by controlling water flow - through the action of a water regulating device on the turbine. Such devices, e.g. spear valves or guide vanes are highly engineered turbine components and are designed for high-efficiency operation. The modern equivalent is the electro-hydraulic governor which uses electrical or electronic means to sense changes in speed but still controls the water flow. The cost of such a governor is often dearer than the cost of the generator at these ratings.

The accepted alternative to the speed governor is the Electronic Load Governor (ELG) which maintains the speed of the set by adjusting an electrical ballast load connected to the generator terminals, maintaining a balance between the total electrical load torque and the hydraulic input torque from the turbine, Fig. 1. In this case, the water flow is kept constant and hence the water regulating device can be dispensed with.

This paper describes the development of a microprocessor based ELG. It goes on to identify limitations in its original performance and to describe improvements in the control algorithm which have led to an advanced performance. Finally, the on-site application of the ELG is described in a Case Study.

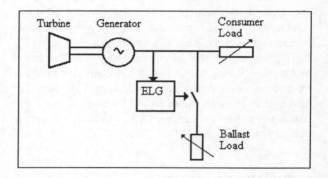

Fig. 1 The load governing principle

TYPES OF ELG

The two most commonly employed techniques used for load governing are the phase delay action and the binary load action.

Phase Delay Action

With this technique, the ballast load comprises a permanently connected single resistive load circuit of magnitude equal to (or slightly greater than) the full load rated output of the generator. As a result of the detection of a change in the consumer load, the firing angle of a power electronic switching device, such as a triac, is adjusted, thus altering the average voltage applied to, and hence the power dissipated by, the ballast load.

As with all power electronic switching of this nature, this technique introduces harmonics onto the electrical system. It is worthwhile to note that these harmonics are continuously present to some extent as long as the ballast load is energised. The presence of these harmonics can cause overheating of electrical equipment connected to the system and of the generator, this is usually counteracted by de-rating of the generating plant, Barnes (3).

Opportunities and Advances in International Power Generation, 18–20th March 1996,
Conference Publication No. 419, © IEE, 1996

Binary Load Action

With this technique, the ballast load is made up from a switched combination of a binary arrangement of separate resistive loads. The load proportion carried by each step is in the ratio 1:2:4 and when switched in sequence, the ballast load thus exhibits a stepped characteristic. The summation of all of the ballast load steps is equal to (or slightly greater than) the rated output power of the generator.

In response to a change in the consumer load, a switching selection is made to connect the appropriate combination of load steps. This switching operation occurs during the transient period only, thereafter full system voltage is applied to the new fraction of the ballast load and hence harmonics are not produced at all by this method in the steady-state. In addition, it is usually the practice to adopt solid-state switching relays which include a zero-voltage switching circuit that reduces the harmonic distortion associated with the transient switching period.

ELG RESEARCH, DESIGN & DEVELOPMENT

A prototype ELG has been designed and assembled which has undergone successful testing on two different Pelton type turbines and on a Francis type turbine on the specially designed test rig in the hydraulics laboratory of Napier University. Three-phase and single-phase versions of the ELG were designed and tested for 50Hz operation, and the electrical output of the test generators ranged from 1kW to 5kW.

Hardware

Considerable research was undertaken into the technical and economic parameters associated with the overall design requirements of an ELG in order to improve on as many of the features associated with ELG design as would be possible. This included study into the extent of the harmonics generated by phase angle control, the phase shift effect caused by integral cycle control, the choice of induction or synchronous type generators, the choice of analogue or digital electronics, the choice of microcomputer and the need for three phase balancing to avoid unbalanced loading of three phase generators. The outcome of this research pointed to a particular design specification which is now summarised.

The ELG was designed for applications in electrical generating systems which are isolated from grid networks and which are fed by three phase, synchronous ac. generators. It was recognised that there may also be applications with single phase generators and an alternative version was prepared for single phase use. Although changes in the consumer load are monitored essentially through frequency measurement, frequency is not in fact calculated by the ELG. Instead it measures and compares the period of the ac voltage waveform. The prototype is designed for 50 Hz systems, however 60 Hz versions would be readily achievable by simple software modification.

The load governing technique adopted is that of binary load switching. As the principle design is for a three phase system, the ballast load in each phase is sectioned into a 1:2:4 format and a current balancing feature is implemented. This requires knowledge of the current in each of the generator lines, achieved through the use of current transformers (CT's).

The design is based on digital electronic circuitry and centres on a microcomputer. A single card microcomputer is used which simplifies its operation and reduces the total number of components, pins and component interfaces. It is expected that this will have a positive effect on the reliability of the unit. The ability to produce ELG units suitable for 50 or 60 Hz systems and either single or three phase simply by modifying the software means that the microelectronic hardware can be identical (excluding the peripheral devices such as switches and CT's). In addition, the flexibility of the software permits solutions to stability control without the addition of analogue filters and flywheels.

The design also minimises the on-site adjustment and operational requirements. In its most basic form, the ELG has only two control variables which need to be set as the unit is first commissioned. Prior knowledge of the ratio of the turbine runaway speed to nominal speed would permit the pre-setting of one of these variables as recorded in a previous paper, (4). The unit is connected only to any two lines at the terminals of the generator. Its internal transformer feeds both the integral power supply circuit and the frequency sensing circuit. The power supply switches on once sufficient voltage is induced on the generator terminals as the generator runs up from standstill. The ELG automatically takes control of the speed of the set from that point until the generator is shutdown.

Software

The software for the microcomputer was written and tested on a development system. The microcomputer used a high-level language, FORTH, and the development system was run on a host PC. Communication between the PC and the microcomputer for program development purposes was via an RS232 serial link. In the target situation, the microcomputer stands alone, the host PC

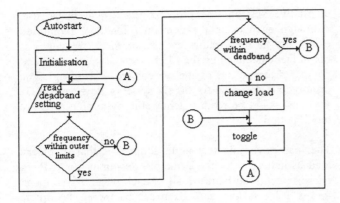

Fig. 2 Flowchart of the Main Load Governing Program

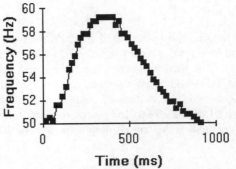

Fig. 3 Typical frequency transient of the
original algorithm

and the serial link are removed and the program is stored in EPROM.

A simplified flowchart of the main load governing program is shown in Fig. 2. Once the necessary auto-start and initialisation procedures are executed, the program enters a continuous loop and reads the first operator-set variable. This is the width of the frequency deadband, resulting from the stepped nature of the ballast load.

The frequency measurement process is handled by an interrupt routine. On detection of a negative going change in polarity of the terminal voltage waveform, a counter is energised. On the repetition of this event (i.e. after one cycle of the ac waveform), the counter is stopped and read, then the value stored in memory. This value is the period of the waveform over that cycle.

The main program compares the most recent reading of the period against values corresponding to the upper and lower limits of the deadband. If the frequency is within the deadband range then no action is taken. If the frequency is outwith the deadband then the ballast load is changed accordingly. An output line is then toggled on or off to enable monitoring of the operation of the ELG. The loop then repeats itself indefinitely until the power supply to the unit is removed.

ORIGINAL ALGORITHM

The original program did not have any knowledge of the actual load condition on the generator and dealt with the change of load by simply adding or removing only one step of ballast load from one phase during each program loop. Therefor in the event that the ballast load was required to change from say zero application to full application, then the minimum number of step changes required was 21 (7 steps in each of the three phases).

Hence the reaction time of the governor was not optimised in any way. With a typical program loop time of 22 ms, this gave rise to a minimum reaction time of 462 ms in the case of a full load rejection. As a result, the maximum frequency, f_{max}, was reaching relatively high values during a transient period caused by such an event.

Using a Pelton turbine as the prime mover, a full load rejection test was performed with the original algorithm controlling the ELG. The resulting frequency vs. time plot for this test is given in Fig. 3. The frequency can be seen to rise above 59 Hz at its peak and the time to return within the deadband, the 'return-time', is in the order of 800 ms. It was thus considered necessary to improve the response of the ELG through the development of an advanced control algorithm.

ADVANCED ALGORITHM

The alternative approach adopted in the course of writing the advanced control algorithm was one of current control. The fundamental aim of load governing is to maintain the generator output *load* constant, and in practice this means maintaining the generator output *current* constant. By measuring the generator current in absolute terms, the extent of changes in the consumer load can be determined and hence the controlling variable, the ballast load, can be adjusted to maintain the generator current at a constant value. This has the advantage that, as the magnitude of the required ballast load change is computed in the algorithm, it can be applied or removed in one program loop.

Software was written by Pearson (5) to implement this control strategy and it was tested in order to provide a comparison with the original program. The results of these tests, on the same turbine, can be seen in Fig. 4 which shows f_{max} reduced by 11% and the 'return-time' reduced by 40% when compared with Fig. 3.

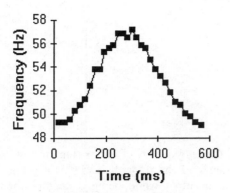

Fig. 4 Frequency transient of the
advanced algorithm

CASE STUDY

The first on-site application of the ELG is at a site being developed in Scotland. The site is located at Ashfield Mill, two miles North of Dunblane on the Allan Water. The existing weir is located at an old cloth dyeing mill site and creates a net head of approximately 6m. Hydro development, providing mechanical power, began at this site in 1925, however the original powerhouse was abandoned in the 1970's. The present site owner and developer is installing two turbine-generator sets to capture the energy available in the water.

The first generating set is now operational and comprises a re-furbished Gilkes turbine and Crompton Parkinson generator. The rated output of this set is 100 kW and it is governed by a traditional mechanical-hydraulic type governor.

It is the second generating set which is governed by the ELG. This is expected to be a Kaplan propeller type turbine running at 750 rev/min and driving, via a belt drive, a 35kVA, 400V, 3-phase synchronous generator. The generator operates in an electrically isolated situation and the electrical energy from the set is primarily used in the residence of the owner for heating and lighting purposes (the consumer load). The ballast load is used for space heating in the powerhouse and in nearby factory units.

This generating set is the subject of a European Commission THERMIE grant to enable on-site demonstration. The CEC Project reference number is HY/329/94/UK. It is anticipated that by the time this paper is presented, the plant will have been installed and that the outcome of the initial tests of the ELG will be available.

CONCLUSIONS

The initial research and development of an Electronic Load Governor suitable for application in isolated situations has been successful. Testing of the original control algorithm has shown up the fact that the methodical application of the binary configuration of the ballast load steps was not ideal. An advanced control algorithm has been developed, specifically designed to address this limitation.

Its implementation results in a small but noticeable improvement (reduction) in the magnitude of the frequency rise on full load rejection and a significant improvement (reduction) in the duration of the transient period itself.

For the Ashfield Mill installation it is the advanced algorithm which has been used in the ELG unit which in turn is the key innovative feature of the European Community THERMIE grant awarded to the site developer.

ACKNOWLEDGEMENT

The author wishes to thank Mr I Robb of Ashfield Mill and Napier University for their support in this project.

REFERENCES

1. Woodward, J.L. and Boys, J.T., 1980, Water Power & Dam Construction, Vol. 32, No 7, p37-39.

2. Wallace, A.R., Henderson, D.S. and Whittington, H.W., 1989, "Capital Cost Modelling of Small Scale Hydro Schemes", Proceedings 24th UPEC, Belfast, UK.

3. Barnes R. 1989. IEE Power Engineering Journal, Vol. 3, No 1, 11-15.

4. Henderson, D.S., 1993, "Recent Developments of an Electronic Load Governor for Micro Hydroelectric Generation", Proceedings HIDROENERGIA 93, Munich, Germany.

5. Pearson, W.N., 1995, "Electronic Load Governor Control Software Modifications", Honours Project Report, Napier University.

STUDIES ON HIGH PERFORMANCE INVERTER GENERATION SYSTEM
OF MICROHYDRAULIC WATER TURBINE

Y.Shimizu, M.Takada, H.Fujiwara, T.Maeda

Mie University, Japan

1. INTRODUCTION

Microhydraulic resource is one of renewable energy resources which has been nearly completely left undeveloped and for that matter, could be an environmentally-sound source of energy since it requires no large scale dam construction and can be co-existing with the surroundings. For the purpose of utilizing renewable energy resources, we have been focusing on the inverter system that makes variable speed control possible for the application to microhydraulic energy and wind energy [Shimizu et al.(1),(2),(3)].

In this paper, the inverter generation system of microhydraulic water turbine with variable speed control in the natural and fluctuating river flow condition is presented and its performance is experimentally evaluated through a number of field tests. Comparison of the inverter system with a conventional fixed-speed control system is also made to show the superiority and advantages of the variable speed control system.

2. INVERTER GENERATION SYSTEM

2.1. Concept of the inverter system

Making continuos and efficient electric power generation possible, variable speed control should be introduced of which concept is schematically illustrated in Figure 1. In a conventional fixed-speed control system which operates only at the rated rotational speed N_r, generated electric power would be significantly decreased from P_r to P_1 as the flow rate gets down to Q_r to Q_1. The developed inverter system which is designed to keep the optimal rotational speed in response to changing flow rate as shown by the curved line in Fig.1 could result in a substantial increase in the total generated electric power especially in the range of low flow rate and low head where a conventional system can not operate.

2.2. Characteristics of induction generator

This microhydraulic generation system is equipped with a 2-kW induction generator. The slip of an induction generator is defined as follows.

$$s = (f_c - f)/f_c \qquad (1)$$

where f_c is the frequency of the exiting current and f is the frequency of the rotor. It works as a generator in case s is negative while being a motor when s is positive. The inverter system automatically controls f_c to keep a negative value of slip throughout the experiments to allow continuous electric power generation.

Figure 2 shows an equivalent circuit of one of the three phases of the induction generator. Neglecting the iron loss that is considered to be quite small, the primary current $\dot{I}_1$ and the secondary current $\dot{I}_2$ will be:

$$\dot{I}_1 = \frac{\{(R_2/s) + j\omega_1 L_{22}\}\dot{V}_1}{\{(R_1 R_2/s) - \omega_1^2(L_{11}L_{22} - L_1^2)\} + j\omega_1\{R_1 L_{22} + (R_2/s)L_{11}\}} \qquad (2)$$

$$\dot{I}_2 = \frac{j\omega_1 L_1 \dot{V}_1}{\{(R_1 R_2/s) - \omega_1^2(L_{11}L_{22} - L_1^2)\} + j\omega_1\{R_1 L_{22} + (R_2/s)L_{11}\}} \qquad (3)$$

where $L_{11} = L_1 - l_1$ and $L_{22} = L_1 + l_2$.

From the above two relations, the generated power per one of three phases of the induction generator will be:

$$P_i = R_1 \dot{I}_1^2 + R_2 \dot{I}_2^2 + \left[\frac{1-s}{s}\right]R_2 \dot{I}_2^2 \qquad (4)$$

where the 1st and the 2nd terms represent the primary and secondary copper losses respectively, and the 3rd term represents the net output power P_m per one phase.

$$P_m = \frac{1-s}{s}R_2 \dot{I}_2^2 \qquad (5)$$

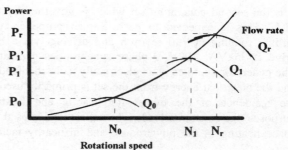

Fig. 1   Power characteristics of an induction generator as a function of the rotational speed.

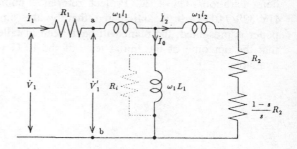

Fig.2   Equivalent circuit of one phase of an induction generator.

Opportunities and Advances in International Power Generation, 18–20th March 1996,
Conference Publication No. 419, © IEE, 1996

Using the relation $P = \omega T$, the total torque of the induction generator of 3 phase and p poles is given by the following relation.

$$T = 3pT = 3p\frac{P_m}{\omega_m} = 3p\frac{1-s}{s\omega_m}R_2 i_2^2 = 3p\frac{R_2}{s\omega_1}i_2^2 \qquad (6)$$

Substituting eq.(3), we obtain:

$$T = 3p\frac{\omega_1 L_1^2 (R_2/s)\dot{V}_1^2}{\{(R_1R_2/s)-\omega_1^2(L_{11}L_{22}-L_1^2)\}^2 + \omega_1^2\{R_1L_{22}+(R_2/s)L_{11}\}^2} \qquad (7)$$

2.3 Control method of the inverter

The inverter developed for our system is a 3-phase voltage-source transistor inverter. Figure 3 shows main components of the inverter unit. The method applied to control the inverter is the Pulth Width Modulation (PWM), which is schematically illustrated in Figure 4. Modulated output waveforms are obtained by comparing a triangular waveform(carrier) and a sinusoidal waveform(signal).

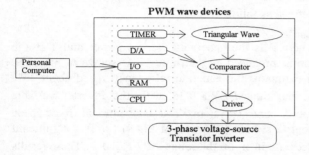

Fig.3   PWM wave devices.

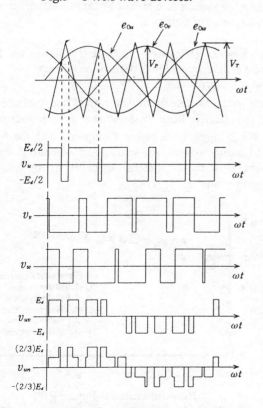

Fig.4   Modulated waveforms by the PWM devices.

3. EXPERIMENTAL APPARATUS

Field experiments were carried out with a 2-kW microhydraulic generation system in Mie University forest. Figure 5 shows the schematic illustration of the microhydraulic generation system. The water turbine used is equipped with blades specially modified to improve its performance, and operates as water from the artificial pond passes through the runner. The maximum static head is measured to be 27 m while the maximum effective head appeared to be about 15 m mainly due to the loss at a nozzle set near the exit to measure flow rate.

For the measurement of flow rate, a nozzle is set near the exit. Effective heads were determined based on static pressures measured by a differential manometer at upstream and downstream the runner. Torque is measured by a torque transducer placed in between the water turbine and a 2-kW induction generator.

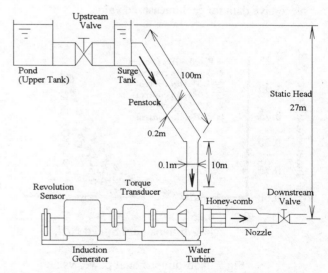

Fig.5   Schematic illustration of the microhydraulic generation system.

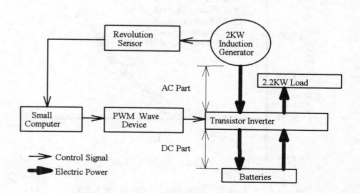

Fig.6   Flow chart of control signal & electric power.

Figure 6 represents a flow chart of control signal and electric power. Alternative current generated by the induction generator is converted into direct current and then is charged to 17 batteries(65Ah each) in series. It was consumed by a 2.2kW heater when overcharged.

4. EXPERIMENTAL METHOD

Experiments were undertaken at various flow rates controlled by the upper and lower valves. To investigate the head effect, tests were carried out at different static heads. During one experiment, slip of the induction generator was kept constant and negative to keep operational in regenerative area. The following values are the applied slip: -1%,-1.5%,-2%,-2.5% and -3%. A conventional generation system in which the primary frequency is fixed at 60Hz was also tested to be compared with the inverter system of variable speed control.

All the data were collected at a sampling frequency of 70Hz, and accumulated by a personal computer. Discussions were made based on the average of 1000 successive data for each measured value.

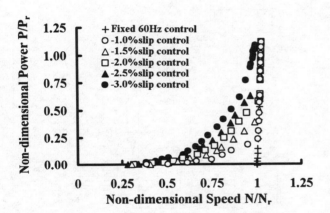

Fig.7 Non-dimensional power vs non-dimensional speed of the rotor.

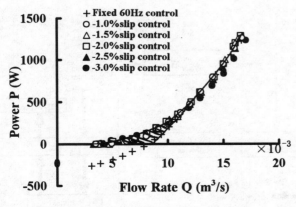

Fig.8 Generated power against various flow rates.

5. RESULTS AND DISCUSSION

5.1 Generated power

Figure 7 shows non-dimensional electric power P/P_r against non-dimensional speed N/N_r. Generated eletric power becomes larger as the slip changes to the bigger in negative(from -1% to -3%) at the same rotational speed. Advantages of the inverter variable speed control system over the fixed speed control system is clearly shown in Figure 8 as the variable speed control system achieved considerable amount of electric power generation in the range of low flow rate where the conventional system could not produce electric power at all. Hence, the shaded area in Figure 8 indicates the improvement in electric power if the inverter system is applied.

5.2 Efficiency of the system

5.2.1 Generator efficiency Generator efficiency η_G is defined as follows:
$$\eta_G = P / T\omega \qquad (8)$$
where P is the generated electric power and T is the torque produced by the runner. As shown in Figure 9, it maintains the generator efficiency η_G of about 70% in the range $P/P_r \geq 0.15$ with the inverter variable speed control system. The conventional fixed-speed control system (+ marks in Fig.9) suffers significant decrease in η_G in the range $P/P_r \leq 0.3$. These results suggest that the inverter system makes it possible to operate the system at a higher generator efficiency over a wider range of flow rate.

5.2.2 System efficiency Figure 10 shows the total system efficiency which is defined as follows:
$$\eta_r = P / \rho g Q H \qquad (9)$$
where ρ is the density of water, Q is the flow rate and H is the effective head. It shows that η_r is higher when the slip is bigger in negative. Compared with the fixed-speed control system, the variable speed control system proved to be more efficient especially in the range of low power area $P/P_r \leq 0.3$.

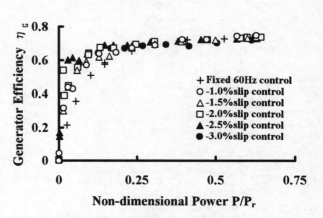

Fig.9 Generator efficiency.

5.3 Comparison of the generated power

Figure 11 shows the generated electric power against the total hydraulic power $\rho g Q H$. The total electric power generated at each value of slip as well as in 60Hz fixed-control system can be obtained by carrying out integration over the range where positive output was gained. Thus, the equations below represent total electric power assuming that the probability distributions of flow rate and effective head are uniform in the integrated range.

Variable speed control:

$$P_v = \int_0^{2400} P_v(E)\,dE \qquad (10)$$

60 Hz fixed-speed control:

$$P_f = \int_{655}^{2400} P_f(E)\,dE \qquad (11)$$

Total generated electric power P_r for variable speed control at each value of slip is shown in Figure 12 in comparison with the total generated electric power P_f of the 60 Hz fixed-speed control system. It is shown that the total generated electric power at each of the applied slip was greater than P_f. The highest improvement is seen at the slip of -2% and -2.5% where the improvement in power became about 22% because the relatively high system efficiencies in the region of $P/P_r \geq 0.3$ contribute to larger electric power at the slip of -2% and -2.5% compared with at -3% slip. It can be concluded that the optimal slip is considered to be -2% and -2.5% slip in this system.

6. CONCLUSIONS

The inverter generation system of variable speed control for a microhydraulic water turbine was developed and experimentally evaluated. The results are summarized as follows.
(1) The inverter system of variable speed control makes it possible to generate electric power in low flow rate and low head where a conventional fixed- speed control system is not operational.
(2) The total generated electric power is greatly improved by about 22% when the inverter system is applied.
(3) The optimal slip in view of maximizing the generated electric power is found to be -2% and -2.5% in this microhydraulic generation system.

ACKNOWLEDGMENT

Authors are very grateful to Prof. Dr. T. HORI, Assoc. Prof. Dr. M. ISHIDA and Mr. N. YAMADA of the Dept. of Electrical & Electronic Eng., Mie University for their help in designing the inverter system.

REFERENCES

1. Shimizu Y, et al. 1986 , "Studies on integration system of multi-micro water turbines set in mountain stream", Proc. ASME 4th Int. Symp. on Hydropower Fluid Machinery, FED-Vol.43, 117

2. Shimizu Y, et al. 1992 , "Studies on computer control of horizontal axis wind turbine and new inverter generation system", Proc. 4th ISROMAC, Vol.B Apr.1-1992, 68

3. Shimizu Y, et al. 1993 , "Studies on wind power generation system with variable rotor speed", Proc. 1st Int. Conf. NESC, 557

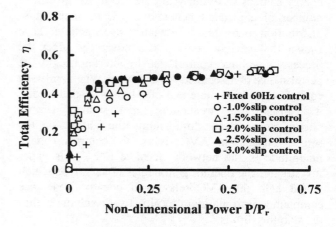

Fig.10 Total system efficiency.

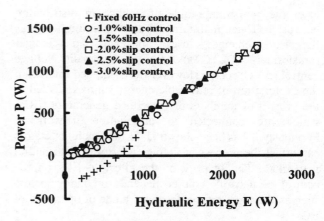

Fig.11 Generated power as a function of hydraulic energy.

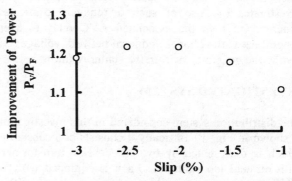

Fig.12 Power improvement at various values of slip.

INVESTIGATION OF THE OPERATING STRATEGIES OF REMOTELY CONNECTED EMBEDDED GENERATORS TO HELP REGULATING LOCAL NETWORK VOLTAGE

S K Salman and F Jiang

The Robert Gordon University, UK

W J S Rogers

Consultant, UK

ABSTRACT

The continued interest in the exploitation of renewable energy sources in recent years has led to the growth of number of embedded generators (EGs) connected to distribution networks. Previous investigations have shown that such generators can adversely affect the process of voltage control due to interference to the performance of associated AVC relays. When an EG is connected at the remote end of a lightly loaded feeder, it causes the weakening and under certain circumstances even diminishing the ability of the source substation AVC relay to control voltage magnitude in the network local to EG. This paper investigates the operating strategies of such EGs which would help the AVC relay with negative reactance compounding to maintain local network voltage within the statutory limits.

INTRODUCTION

Over the years magnitude of voltage in distribution network has been maintained within operating limits as the load changes, using on-load tap changer transformers (OLTCTs) with an automatic voltage control (AVC) relay. However the continued interest in the exploitation of renewable energy sources has led to the growth of numbers of embedded generators (EGs) which are connected to distribution networks. Previously it has been shown [1-5] that such generators can affect the process of voltage control of distribution networks. The severity of the effect of EGs on the control of network voltage depends on many factors such as the type, rating, location, mode of operation of EGs.

This paper deals mainly with the effect of EGs which are connected at a distance from the source substation where the AVC relay is located. The paper particularly investigates the use of such a remotely connected generator to help the associated AVC relay to keep general network voltages and local network voltage, i.e. nearby the generator, within the statutory limits.

INVESTIGATED SYSTEM

The distribution system considered in this investigation is shown in Fig. 1. Basically it consists of a substation which is represented by two 33kV/11kV transformers with on-load tap changer. Each is equipped with an AVC relay to maintain the connection voltages within the statutory limits. The substation is connected to a grid through 33kV line on one side while it feeds 11kV feeders on the other side. The details of only one feeder is shown while all other feeders are lumped as Load 1.

VOLTAGE CONTROL OF A FEEDER HAVING EMBEDDED GENERATOR(S)

It is important that voltage magnitude along a feeder is maintained at almost constant value.

(i)   Without Embedded Generator(s)

In the absence of embedded generator connected to the feeder, this is usually achieved by combining the following two actions.

(a)   Any change in the magnitude of source voltage caused by load variation is offset automatically by the AVC relay installed on the transformer. The relay continuously monitors the voltage magnitude at bus No 3, as follows:

$$V_r = V_3 - I_L \left(R_{avc} - j\, X_{avc} \right) \qquad \ldots (1)$$

where V_r and V_3 are the predicted voltage by the relay and the voltage at bus No 3 respectively, I_L is the load current throughout the 33kV/11kV transformer and R_{avc} and $-X_{avc}$ are the resistive and negative reactance compounding respectively of the relay. These two relay parameters may be adjusted for constant voltage or to model the 11kV feeder.

When the voltage predicted by the relay goes beyond its bandwidth setting, the relay operates the on-load tap changer in such away to bring voltage magnitude back to be within the limits.

(b)   Most customers are connected to 11kV feeders via 11kV/LV (where LV is equal to 230V -6% to 230V + 10% per phase) transformers which are normally equipped with off-load tap changers, the position of such tap changers should be selected such that voltage magnitudes at the consumers' premises are always maintained within the statutory limits as the total connected load on the source substation varies between its minimum and maximum values. For operating flexibility this should be achieved for alternative supply arrangement without having to change the tap position of any of the HV/LV transformers along the 11kV feeder under consideration.

Opportunities and Advances in International Power Generation, 18–20th March 1996, Conference Publication No. 419, © IEE, 1996

(ii)   With embedded generator(s)

When an embedded generator is connected to an 11kV feeder the ability of the associated AVC relay to maintain voltage magnitude along the feeder within the statutory limits will be affected by various degrees. This depends on factors such as the rating, location, mode of operation of the connected embedded generator(s) and the load condition. Considering the worst case where the generator is connected to the remote end of an existing lightly loaded 11kV feeder. If the rating of the generator is greater than the load connected to the feeder at the location of the generator, the power has to be send towards the source substation. This will lead to a voltage difference between that of the bus at which the generator is connected i.e. bus 5 (see Fig. 1) and 11kV bus 3 at the source substation. Taking V_5 (see Fig. 1) as reference and neglecting the load connected at bus No 4, the voltage difference between V_3 and V_5 is given as follows:

$$\Delta V = \Delta V_X + j\Delta V_y \qquad ... (2)$$

where ΔV_x and ΔV_y are the horizontal and vertical component of voltage difference between buses 5 and 3. Their magnitudes can be found in terms of the net active and reactive powers P and Q respectively supplied by the generator to the source substation and the resistance R and reactance X of the feeder connecting the generator and the substation as follows [6]:

$$\Delta V_x = \frac{RP + XQ}{V_3}$$

$$\Delta V_y = \frac{XP - RQ}{V_3}$$

$$\therefore \quad V_5 = V_3 + \frac{RP + XQ}{V_3} + j\frac{XP - RQ}{V_3} \qquad ... (3)$$

However the difference between V_5 and V_3 can be approximated by neglecting the j-term as shown by eqn (4):

$$V_5 - V_3 = \frac{RP + XQ}{V_3} \qquad ... (4)$$

It will be noted that generally the resistance, R, of an 11kV feeder is usually greater than its reactance, X. Therefore the voltage drop along the feeder between the EG and the source substation is determined by both active and reactive power flows through the feeder as shown in eqn (4). It is evident therefore that when RP + XQ > 0, voltage magnitude at bus No 3 will be less than that at bus 5. However if RP + XQ < 0, the voltage magnitude at bus 3 becomes greater than that at bus 5. Table 1 shows all possible operating conditions.

Table 1: The relationship between voltage magnitudes at buses 5 and 3 (see Fig 1) under the different load flow conditions.

	The sign of net power delivered by EG to the Grid at the S/S		RP+XQ	voltage relationship	The possible operating conditions
	$P=P_G - (P_L + P_{loss})$	$Q=Q_G -(Q_L + Q_{loss})$			
case 1	$P \leq 0$	$Q \leq 0$	≤ 0	$V_5 \leq V_3$	a) Without EG(s) b) With an EG having a rating smaller than the local load.
case 2	$P \geq 0$	$Q \leq 0$	≤ 0	$V_5 \leq V_3$	The generator operates at a leading power factor, ie, absorbs reactive power, and its active power is greater than the local feeder load.
case 3	$P \geq 0$	$Q \geq 0$	≥ 0	$V_5 \geq V_3$	The generator operates at a lagging power factor, ie, supplies reactive power, and both its active and reactive powers are higher than those of the feeder load plus its losses.

182

where P_{loss}, Q_{loss} are the active and reactive power losses respectively over the feeder between the EG and source substation.

Case 1 of this table is related to the situation where either there is no EG is connected to the system or an EG connected at bus No 5 but with a rating less than that of the local load. Under this condition the voltage magnitude of bus 5 can still be regulated within the statutory limits by the source substation OLTCTs. However in cases 2 and 3 which are related to an EG with a rating greater than that of the local load, OLTCTs can not properly control voltage magnitude of the busbar local to the EG, ie bus 5. Because this bus becomes a source bus and therefore its voltage is mainly determined by the operating mode of the generator connected to it. This will demonstrated under the section related to results.

Equation (4) and table 1 show that the difference between V_5 and V_3 for a specific generator power P depends on the relative values of R and X of the feeder and more importantly on the reactive power supplied (Q has positive value) or absorbed (Q has negative value) by the generator. Consequently the operating strategy of an embedded generator as far as voltage control is concerned is decided mainly by the value of supplied or absorbed reactive power compared to generator output.

SAMPLE RESULTS

The 33kV/11kV distribution network shown in Fig 1 is used in this investigation. This has the extreme circuit lengths to demonstrate the issues under consideration. The setting of the off-load tap changers of transformers T1, T2 and T3 to maintain the voltage at buses 6, 7 and 8 within the statutory limits as the total load connected to the source substation varies from zero to its maximum value is found to correspond to a voltage ratio of 0.975/1.0. This is based on using the standard settings for the AVC relay. The standard settings are defined as those settings which ensure that the magnitude of the predicted voltage $|V_r|$ to be the same as that of bus 3 i.e. $|V_3|$. For the network under consideration these settings correspond to $R_{avc} = 1\%$ and $X_{avc} = -5\%$.

In this investigation it is assumed that an embedded generator is connected at the remote end of the 11kV feeder, i.e., at bus No 5. The maximum rating of the generator is mainly decided by the length and the cross sectional area of the feeder [5,7]. For the feeder under consideration the maximum rating of the generator which can be connected is found to be 300 KVA. Since the load connected to the feeder under consideration is small compared with that of the generator, the latter is assumed to send its power partially along the feeder

towards the substation. The effect of various operating strategies for the generator on the local (within the vicinity of the generator) voltage profile have been examined. This includes operating the generator in constant power factor mode and in constant voltage mode.

Constant power factor mode operation

The effect of operating a remotely connected embedded generator in a constant power factor mode on the voltage profile of the associated 11kV feeder has been investigated. The power factors that have been considered are 0.98 lagging, 0.85 lagging and 0.98 leading. The lagging and leading power factor conditions are corresponding to the generator supplying and absorbing reactive power respectively. This investigation has shown that operating the generator at a lagging power factor generally tends to raise local voltage magnitude. On the other hand operating the generator at leading power factors tends to reduce local voltage magnitude. However, it has been found that to bring the voltage sufficiently low to ensure that its magnitude at customer sites is within statutory limits the generator is required to be operated at a low leading power factor which means that the generator under this condition is required to absorbed large reactive power.

Fig. 2 and 3 show the influence of the AVC relay at the source substation on the voltage profile of buses 4 - 8. This corresponds to the connection at bus No 5 of a 500 KVA generator operating at constant power factors of 0.98 (lead) and 0.85 (lag) conditions respectively.

Constant Voltage Control Mode

The previous study is repeated with the magnitude of the generator terminal voltage is held constant under the action of the generator AVR and/or its associated capacitive/inductive var injection system. The voltage profile along the 11kV feeder is then examined by keeping generator terminal voltage magnitude constant at 1.03 pu, 1.01 pu and 0.99 pu. In each of these three cases the base voltage at bus No 3 which is controlled by the AVC relay is assumed to be kept at 1.01 pu. This investigation has also examined the effect of the type of interfacing circuit between the generator and the system. This includes direct connection and the connection of the generator via a transformer.

Fig. 4 shows the voltage profile of the AVC relay and buses 3-8 for the case when the generator is connected via a transformer to bus No 5. This corresponds to a generator terminal voltage of 1.01 pu, i.e. at the same value as the AVC relay base voltage. Fig. 5 on the other hand shows the corresponding variation of reactive power which should be handled by the generator corresponding to generator terminal voltages of 1.03 pu, 1.01 pu and 0.99 pu respectively. It can be seen that the generator is required to handle a wide band of reactive power which is equal to approximately 420 KVAR as the load connected to the source substation is varying from zero to its full load and vice versa.

This investigation has also showed that maintaining the terminal voltage of the generator at 1.03 pu, i.e. at a value higher than that of the base voltage at bus 3 causes the reactive power cycle curve to be pushed vertically upwards without appreciably changing its overall band value. On the other hand the same power cycle curve is caused to be shifted downwards when the voltage magnitude of the generator is maintained at values less than that corresponding to the base voltage value of bus 3. It is found that this is applicable for both types of generator interfacing.

CONCLUSION

The effect the operating strategies of a remotely connected generator have on the voltage profile of a long associated 11kV feeder has been investigated. Two operating strategies have been considered. These are constant power factor mode and constant voltage mode of operations. This investigation has shown: (i) the generator generally increases local voltage magnitudes when it is operated at lagging power factor conditions, (ii) local voltage magnitude can be reduced by operating the generator with leading power factor conditions. However to maintain voltage magnitude within the statutory limits, it is required to operate the generator at a low (leading) power factor condition, i.e. the generator and its associated capacitive/inductive var system are required at times to absorb large amount of reactive power, and (iii) local voltage can be best controlled if the generator is operated at constant voltage mode. However this type of operation may place heavy demand on the generator to supply or absorb reactive power to/from the network particularly when the generator is connected directly to the system. The amount of reactive power which has to be handled by the generator is found to be considerably reduced if the generator is connected to the system via a transformer and when source voltages are not axcessive.

ACKNOWLEDGEMENT

The authors would like to thank RGU for providing the facilities to prepare this paper and Manweb plc for its continued support to the work on embedded generation.

REFERENCES

[1]   Kirkham, H and Das, R 1984, "Effect of voltage control in utility interactive dispersed storage and generation systems", IEEE Trans. Power Appar. & Syst., 103, 2277-82.

[2]   Salman, S K, Jiang, F and Rogers, W J S, 1992, "The impact of addition of private generator on the voltage control of an 11kV network", 27th UPEC Conf, 2, 530-33.

[3]   Salman, S.K., Jiang, F. and Rogers, W J S, 1993, "The effect of private generators on the voltage control of 11kV networks and on the operation of certain protective relays", Athens Power Tech: IEEE/NTUA Inter. Conf. on Modern Power Systems, 2, 591-595.

[4]   Salman, S K and Jiang, F, 1993, "Comparison between the effects of embedded synchronous and induction generators on the voltage control of a distribution network", 28th UPEC Conf, 1, 401- 4

[5]   Salman, S K, Jiang, F and Rogers, W J S, 1993, "Effects of wind power generators on the voltage control of utility's distribution networks", Inter. Conf. on renewable energy - Clean power 2001, IEE Publication No 385, 196-201.

[6]   Weedy, B M, 1987, "Electric Power Systems", John Wiley, Norwich, UK

[7]   Taylor, E and Boal, G A, 1966, "Electric Power Distribution 415-33kV", Edward Arnold, London, UK

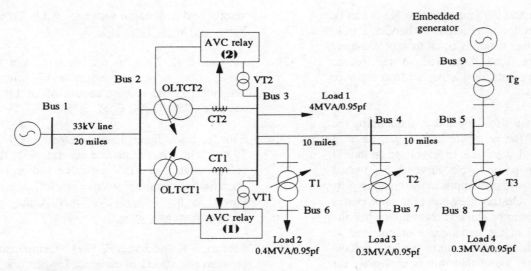

Fig.1 The simulated 33kV/11kV distribution system

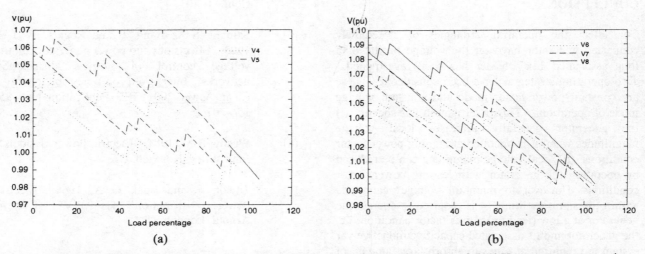

(a)

(b)

Fig. 2 Voltage profile of (a) buses 4 - 5 and (b) buses 6 - 8 due to connection of 500kVA generator operating
at constant power factor of 0.98(lag)

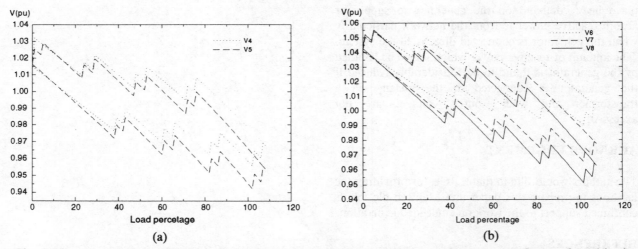

(a)

(b)

Fig. 3 The same results as Fig 2 but the generator is assumed to operate at a constant power factor of 0.85(lead)

185

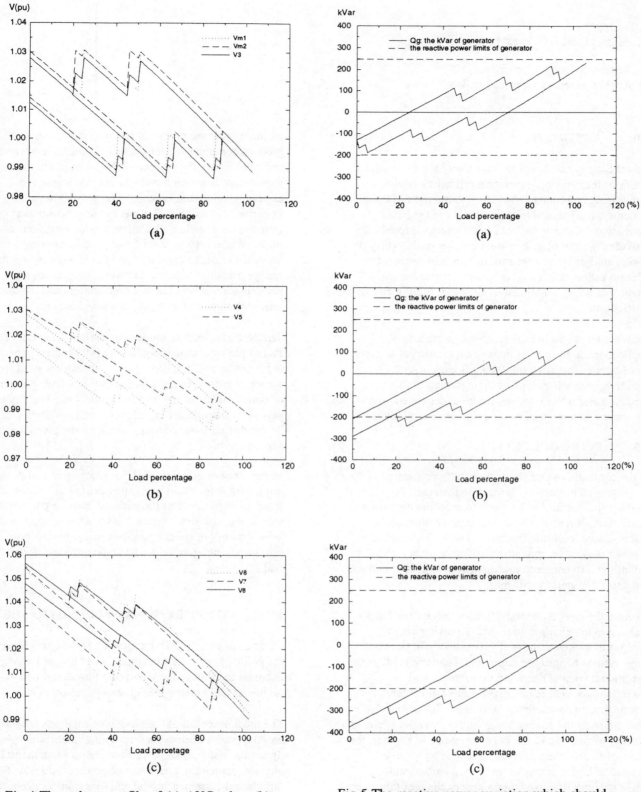

Fig.4 The voltage profile of (a) AVC relay, (b) buses 4 - 5 and (c) buses 6 - 8 due to the operation of generator at constant voltage control mode which corrensponds to genera -tor terminal voltage of 1.01pu

Fig.5 The reactive power variation which should be handled by the generator at terminal voltages of (a) 1.03pu, (b) 1.01pu and (c) 0.99pu

EFFICIENCY AND EMISSIONS: COST EFFECTIVE MODELLING FOR PLANT PERFORMANCE IMPROVEMENT

C. F. King*
R. J. Ettema and J. C. Paul**

* Airflow Sciences Corporation, UK
** Airflow Sciences Corporation, US

1.0 ABSTRACT

Power Station managers are faced with the two major issues of maximising generation efficiency whilst minimising emissions. Numerical simulation of boiler systems has advanced to the point where the best 3 dimensional Computational Fluid Dynamic models can provide a wealth of information and be used to carry out design studies for the optimisation or improvement of combustion efficiency, emissions, corrosion and erosion damage, heat transfer, slagging, fouling and flow distribution.

Drawing on the authors' extensive experiences of applications in the power generation industry, the paper demonstrates that such modelling is extremely cost effective, both when applied to upgrading existing hardware and when used to provide independent and objective assessment of new equipment installations.

2.0 INTRODUCTION

In the competitive and environmentally concerned 1990s, two issues face every power station manager. The first is whether the station can be made to meet the emissions legislation. If it can't do so within an economically viable frame work the station will shut. The second is whether the station can generate electricity at a cost that will allow it to compete successfully for supply contracts. If it can't, the station will shut.

Meeting the environmental and commercial challenges means making changes to plant. Knowing which changes to make and how effective they will be is not easy. Some changes are incremental and are made at the operational level. Others are more major and discontinuous. These can require substantial capital investment, and sometimes involve modifications that have the potential for coupling improvement in one area with significant reduction in performance in another, for example improvements made to reduce NO_x have resulted in higher levels of unburned carbon and increased rates of corrosion on heat exchanger tubes due to reducing atmospheres.

Knowing what the outcome will be before embarking on change gives Station Managers a lot of comfort. Identifying the potential savings or increased revenue before any capital is committed allows choices to be made between competing priorities. This kind of prediction requires understanding of a type that can only come from detailed modelling of the specific plant geometry and the proposed changes to it. The best modern computer modelling techniques have advanced to the point where the correlation between prediction and experiment is within the limits of accuracy of the experimental measurements. The authors have experience of using such tools in many power stations to improve combustion efficiency, reduce emissions and improve heat rate, to identify ways of improving fuel and air distribution, to examine the effects of changing fuel, to reduce pressure losses in ductwork, to enhance the performance of electrostatic precipitators and mechanical dust collectors and to reduce maintenance costs.

The cost of computer modelling is often no greater than that of physical modelling but the information generated is far greater. Computer models can also be used for design purposes, quickly and effectively trying a number of options to allow the best to be selected. The payback period on such modelling is short, well within the 12 to 24 months necessary. Many studies have given a much shorter payback.

In the sections below a brief introduction to computer modelling is followed by examples of applications taken from the experience of the authors' company in working with power utilities on nearly 100 different applications. Where names of specific utilities companies or power stations are not given the data is nevertheless drawn from real situations.

3.0. COMPUTATIONAL FLUID DYNAMICS

Computational Fluid Dynamics (CFD) has developed to the point where numerical models can be used, to examine the flows in any part of a power station. A review of CFD methods is given by Nelson et al[1].

The most powerful of the modelling methods are those based on three-dimensional computational fluid dynamics. CFD models are founded on fundamental physical principles, and can thus predict fluid flow (gases, liquids and entrained solids) and heat transfer properties within boilers under virtually any operating conditions. With the recent addition of combustion and turbulence submodels, these methods also provide a full representation of chemically reacting flows throughout the furnace. Thus, combustion air, burner, fuel handling, heat exchanger and flue gas cleaning systems can all be analyzed in detail.

Opportunities and Advances in International Power Generation, 18–20th March 1996,
Conference Publication No. 419, © IEE, 1996

In the authors' experience, the application of CFD to power station steam generation systems has the capability to tackle a wide range of technical and economic issues within a power plant. A list of economic issues associated with various components of fossil-fuelled power plants is presented in Table 1. Each economic issue is associated with a specific technical subject area that has been evaluated using CFD modelling techniques. Numerical models can be used to define the impact of alternative design and operating conditions on boiler performance and efficiency. The economic impact on plant performance can be substantial, as will be illustrated in the remaining sections of this paper. The effect that numerical models can have on power plant economics readily justifies their use.

4.0. COST : BENEFIT FOR SUBSYSTEM PERFORMANCE IMPROVEMENTS

A fossil fuelled power station is a collection of subsystems. Poor performance in one subsystem affects the rest because of the close inter-relationship between them. The knock on effect of off design performance in one subsystem can seriously affect the overall plant economics. The costs to the plant occur as operations and maintenance (O & M) expenses, regulatory penalties, capital expenses or lost revenue from reduced plant output.

While O & M costs are unavoidable and are therefore budgeted items, poor system performance can quickly consume a "reasonable" budget and lead to major overruns. The same is true of unanticipated capital expenses. In countries such as the USA, regulatory penalties for excess emissions can be substantial. They are avoidable by close control using sophisticated emission control systems but the avoidance can require reduced output due to derating and thus incur lost revenue costs, see Figure 1. Poor combustion results in unburned fuel and an increased Net Plant Heat Rate (NPHR). Thus the amount of fuel energy (BTUs) required to generate one kilowatt hour of saleable electrical energy is increased. Such increases in NPHR have the immediate effect of raising fuel costs and are generally indicative of other performance problems as showed by Speckman et al (reference 2) in a 1986 review of Pacific Gas and Electric Company. The importance of NPHR to annual coal costs is shown in Figure 2. Unburned carbon in ash not only has impact on the coal costs but also leads to problems in electrostatic precipitator efficiency due to lowered fly ash resistivity and can make the fly ash unsaleable. For many plants, the economics of the plant are crucially affected by the difference between paying to dump high carbon content fly ash and being paid for low carbon content fly ash as concrete filler. With increasing numbers of plants offering fly ash to the market in UK the concrete manufacturers are now demanding ever lower levels of carbon in ash.

4.1. Coal Pulverizers

Improvements in pulverizer performance offer substantial savings in fuel costs. Finer grind and tighter size bands have been shown to substantially reduce unburned carbon. The Tennessee Valley Authority has reported that pulverizer improvements in one of their units has reduced carbon in ash from 10-11% to 4-5% This has resulted in a £50K per month savings in direct fuel costs. Fuel costs due to carbon in ash are summarised in Figure 3. Even though pulverizers tend to be geometrically complex, air flow and individual coal particle paths can be computed. The results are utilized to improve size classification and reduce particle residence time and system pressure losses within a pulverizer. While the studies can require an investment as high at £70K, a reduction of only 2% in carbon in ash at a typical 6500MW_e plant would provide a 10 month payback.

4.2. Non-Uniform Combustion

Consider the problem of non-uniform combustion of pulverized fuel (PF). Localised fuel-rich and fuel-lean regions occur within the furnace when air and coal are unevenly distributed between the burners. This condition many exist even when indicators such as flue gas O_2 levels read "normal"[4]. Non-uniform combustion often results in increased emissions and in high slagging rates thus increasing NPHR with penalties as outlined above.

A midwestern US utility operating a 360MW_e cyclone fired unit recently estimated the annual cost of slag removal at £40-60K plus shutdown and lost revenue costs. Conservatively assume that the utility was able to limit its outage time for cleaning to only 48 hours over the year. Even if the utility was able to make up the power output by firing a slightly less efficient standby unit (0.05p/k Wh additional generation cost), the lost revenue would amount to £60K annually. Thus, the total annual cost of slag removal plus higher generation costs could run from £100-130K.

A numerical study to restore the primary and secondary air balance in this particular unit would cost in the £40-60K range. The recommended modifications resulting from these studies are simple in nature. With a conservative assumption of £10K for fabrication and installation, and only a 50% decrease in slagging rate, the payback period for the study would still be just 12-16 months.

Some utilities routinely deal with wall fired PF unit slagging problems by operating below the unit design load. Derates of 5% are common for this purpose. A Michigan utility currently operating a 630MW_e unit at 600 MW_e estimates their cost in lost revenue at £3000 per day. In addition, the unit must be taken off line every 4 to 6 months for deslagging operations. Annual costs due to the derate alone are in excess of £1.0 million assuming an 85% unit availability. If numerical studies to balance primary air/coal and to balance secondary air are completed at a cost of £50K each, an output gain of just 3MW_e would provide a one year payback.

4.3. NO$_x$ Control Systems

In order to comply with increasingly stringent pollution regulations, many power companies must purchase new equipment to limit plant emissions. Fines for excess NO_x emissions in the USA for example can be quite severe as indicated by the graph in Figure 4[5].

Options for reducing NO_x emissions include Overfire Air (OFA), Natural Gas Reburning (NGR), Selective Catalytic Reduction (SCR), Selective Non-catalytic Reduction (SNCR), and Low NO_x Burner (LNB) systems. Faced with such major capital expenditures, companies have been struggling to find the most cost effective solution for their unique situations. Even though few other truly scientific evaluation methods are available, many of these companies are understandably reluctant to add to their costs by conducting numerical simulation analyses.

In relation to the capital costs of NO_x control measures, however, the simulation expense is minuscule. Average retrofit costs computed from several sources,[6, 7, 8] are plotted Figure 5. If the numerical analyses are considered as "insurance" on proper performance of the NO_x reduction system, the costs are easily justifiable against the excess emissions penalty.

Consider a 600 MW_e wall fired PC unit which installs the low costs OFA option to lower emissions from 0.6lb/MBTU to less than the 0.50lb/MBTU limit. If the modification succeeds the £3.5million investment offers a payback in about 1.7 years again the £2 million (first year) potential EPA fine If a full numerical combustion study is included at an additional cost of £160K the payback period is increased by less than one month. In addition, it is possible that the modifications resulting from the numerical analysis could eliminate the need for the OFA system altogether.

At the other end of the cost spectrum, SCR systems can be extremely effective provided the distribution of ammonia in the flue gas, and the distribution of flue gas through the catalyst is reasonably uniform. Poor gas distribution can also reduce catalyst life. Since the catalyst is the single largest contributor to SCR cost[9] non-uniform gas flow can have significant long term O & M cost penalties. Numerical flow distribution studies to ensure correct flow distribution are relatively inexpensive, ranging from £10-50K. The change in the payback period from adding such a study to the capital cost of a SCR system is virtually negligible.

4.4. SO₂ Control Systems

The other emissions issue with a major economic impact is sulphur dioxide (SO_2) control. Various technologies are available ranging in price from £50/kW for dry injection systems to as much as £200/kW for regenerable systems[10]. Capital costs are thus measured in the tens of millions of pounds. Wet scrubbers, for example have a high SO_2 removal efficiency, and provide some cost savings by allowing the sale of excess emissions allowances in the USA and by making it possible to burn a cheaper, high sulphur coal. Wet scrubbers work by injecting a liquid reactant into the flue gas stream as it passes through a large tank. As with SCR systems, the scrubbing efficiency can be compromised by non-uniform distribution of flue gas in the tank. Without even considering the capital cots for the system or the EPA fines, potential savings from burning less expensive coals justify the numerical study cost. If scrubber efficiency is improved just enough to allow a 0.2% increase in coal sulphur content, a 600MW_e plant could save approximately £1.5 million in annual fuel costs The payback period for a £30K numerical study would thus be 10 days!

On the other hand, switching to coals with a lower sulphur content offers a means to avoid the high costs associated with scrubbing systems. Plants originally designed to burn a variety of coals may be able to switch with only minor modification. However, other plants specifically designed for high sulphur coals may require major rework on coal handling and storage, primary air (conveying), slagging, soot blowing, steam attemperation, and ash removal systems[11]. Thus, total costs due to fuel switching will vary widely, but can easily reach in to the millions of pounds. Given an ultimate analysis of the new coal, boiler combustion simulations can predict the resulting SO_2 concentrations as well as other products of combustion and carbon in ash. At a cost of £60-150K payback can be expected in a matter of months when set against the potential capital and O & M fuel switching costs.

4.5. Steam Tube Banks

Fly ash present in coal fired units frequently leads to fouling and erosion in steam tube banks, particularly in economisers where the tubes are closely spaced. This problem is exacerbated in units switching to lower sulphur, high ash coals. A utility burning low sulphur Powder River Basin coal estimates that removal of accumulated ash from a 138 MW_e unit costs £30-50K every 2-3 years[12]. For cases of severe erosion, entire tube banks may have to be replaced. A Pennsylvania utility recently completed a reheater and economizer replacement at a cost of £5 and £2.5 million, respectively. Numerical analyses of back pass conditions including ash accumulation and erosion can be typically completed at costs ranging from £20-50K. If the analysis saves a single £2.5 million tube replacement over a 30 year plant life, an annual saving of £90K (constant-cost basis) would be realised.

4.6. Air Preheaters

Fly ash also poses fouling problems to air preheaters. In regenerative (Ljungström and Rothemühle) preheaters, ash particles larger than approximately 1mm (0.04) inch can be trapped between the heat transfer plates leading to eventual forced outages for cleaning. Tubular air preheaters can also foul when ash becomes trapped in sulphuric acid condensing on the tube surfaces. The Dayton Power & Light Company was experiencing forced outages every 2 to 3 months in a 600 MW_e unit due to fouling of the heat transfer plates. With water wash costs of about £12K per outage, cleaning expenses alone ranged from £50 - £80K annually. A numerical design study of this unit reduced the number of required cleanings to one per year which is completed during the scheduled annual maintenance outage. Given a £20-40K study cost and a £25K fabrication and installation budget, payback against the cleaning expense would be from 0.8 to 2.4 years.

4.7.   Electrostatic Precipitators

As ageing electrostatic precipitator units are updated to meet current emissions standards, rebuilds are becoming more frequent. With rebuild costs of approximately £120/m^2 of collection area [13] (materials plus labour), this service can run up to £1M or more for each unit. The new electrical components will function at maximum efficiency only if the gas flows uniformly through the collection plates. Local areas of high velocity gas can sweep significant amounts of dust through the system and increase stack opacity.

An eastern utility operates a unit that is frequently derated upto 25MW$_e$ due to high opacity. When demand is high during the derate, the utility must buy power at £50/MW$_e$h above their normal generation cost. If this condition occurs for a total of 7 days during the course of a year, the cost of buying power would pay for a £175K precipitator rebuild plus a £30K numerical design study to ensure uniform flow.

For a Michigan utility operating a 40MW$_e$ plant, opacity exceedance forced a continuous maximum load of 33 MW$_e$.[14]. A numerical simulation of the precipitator was utilized to design new inlet and exit perforated plates and thereby create a uniform flow field through the collection region. No electrical changes were made. Following the installation of the modification, full load opacity was reduced by over 50%. Full load operation was then restored with no opacity penalties or derate costs.

4.8.   Continuous Emissions Monitors

The location of Continuous Emission Monitoring System (CEMS) sensors is known to affect the accuracy of the flow volume readings.
Certain flow conditions such as cyclonic flow in the stack can make accurate readings impossible. If the monitor gives a reading that is too high, a US utility will be forced to use more emissions allowances than necessary. US Federal law prohibits monitors being biased downward. A utility than emits 100,000 tons of SO$_2$ annually can lose over 5,000 allowances. At a cost of £100 each, the value of the wasted allowances would be in excess of £500K. At an average cost of £30K per breeching duct, 15 different single duct stacks could be corrected for the annual allowance cost.

4.9.   Duct Pressure Losses

Pressure losses in ductwork can have a high cost in terms of additional power consumed by ID and FD fans or by lost power to gas turbines. A California utility operates

four gas turbine units that are equipped with Heat Recovery Steam Generators (HRSG). High pressure losses through the HRSG and ducts leading to the stack cost the utility £13K per year per inch of water loss for each of the four units. A £25K study to design minimal pressure loss ductwork before construction would be paid back in one year with just a ½ inch pressure recovery.

A UK power station was forced to derate in certain weather conditions due to ID fan limitation. As part of an £80K study to design ESP and ductwork upgrades, ductwork design modifications were predicted to yield a 6% reduction in flue gas path pressure loss. The reduced fan power alone will repay the investment within 12 months, while the potential for increased revenue from avoided derating could repay the study cost within weeks.

5.0.   CONCLUSION

Modern fossil-fuelled power plants have complex interacting subsystems, poor performance in any one of which can have severe economic ramifications. Station managers are faced with the twin goals of minimising emissions and maximising revenue. In order to perform this task they need cost effective tools that can deliver an understanding of the necessary changes to the plant. Numerical modelling of plant provides accurate evaluations of harsh and hazardous environments. CFD based 3-D numerical models provide the versatility to analyse the boiler systems in any operational regime and allow design studies to be performed to correct deficiencies.

Examples drawn from the authors' experience of almost 100 applications, given in the paper, demonstrate that numerical modelling provides answers which permit cost effective plant modifications. While the cost of computer modelling is substantial, the payback period has been shown to be well within the profitable period of half the remaining life-time of the modified subsystem [15]. In many instances the payback period was substantially less than one year.

Thus when planning budgets for modification and development of plant, funding should be allowed for cost effective, independent assessment of existing plant performance and the potential for future improvement by upgrade, replacement or addition.

6.0.   REFERENCES

1. Nelson, R.K., Franklin, J.D., Scherer, B.E., "Prediction of NO_x Generation in Coal Fired Boilers", Power-Gen Americas 94, Orlando,Florida, December, 1994; Airflow Sciences Corporation Report R-94-ASC-004, Livonia, Michigan, September 1994.

2. Speckman, Bernard M., and Smith, Morton J, "Heat Rate Deviations as   a Meaningful Performance Measure" Public Utilities Fortnightly, April 2, 1987 pp-34-40

3. Moates. R.L., Tennessee Valley Authority, Telephone conversation, January 25, 1995.

4. Thompson, R. and Dyas, B., "The Role of Combustion Diagnostics in Coal Quality Impact and NO_x Emissions Field Test Programs," EPRI Conference : Effect of Coal Quality on Power Plants, August 17-19, 1994

5. Moyer, Craig A., and Francis, Michael A., Clean Air Act Handbook,Second Edition, pp4-15, Deerfield, Illinois; Clark BoardmanCallaghan. 1992.

6. Smith, Douglas J., "Low-NO_x burners Lead Technologies to Meet CAA's Title IV", Power Engineering, June 1993, pp 40-42

7. Eskinazi, D. et. al., "Retrofit NO_x Controls for Utility Boilers, A Synthesis of Technologies, Issues and CAAA Ozone Attainment Legislation," EPRI Report TR-10212, October, 1993

8. Castaldini, Carlo, et al, "NO_x Reductions and Controls for Utility Boilers in the NESCAUM Region", Acurex Environmental Corporation, 1991.

9. Herwitz, Howard, et. al., "Cost Effective Solutions for Reduction of NO_x Emissions from Coal Fired Utility Boilers," Power-Gen Americas '93, Dallas, Texas, November, 1993

10. Helfritch, D.J., and Kumar, K.S., "Comparative Economics of SO_2 Control Technologies, Proceedings of the American Power Conference Vol 55 pt 2, Chicago, Illinois, 1993

11. Pavlish, John H., et. al., "Using the CQUIM" To Evaluate Switching to Western Low-Sulphur Coals," Paper presented at the Engineering Foundation Conference of Coal Blending and Switchings of Low-Sulfur Western Coals, Snowbird, Utah, September - October, 1993

12. Bleakley, Kevin, Northern Indiana Public Service Company, Conversation, November 14, 1994

      , James, and Phelan, Brian, PrecipTech, Inc.,
      one Conversation, February 10, 1995

14. Peterson, Dennis, Upper Peninsula Power Company, Letter to Airflow Sciences Corporation, October 29, 1991

15. Payne, F. William, Efficient Boiler Operations Sourcebook 3rd Ed. Fairmont Press, Lilburn, GA, 1991

SUBSYSTEM	TECHNICAL ISSUES	ECONOMIC ISSUE(S)
Primary Air Conduits	Air distribution between burners	Emissions, slagging, and heat rate increase due to inefficient combustion
	Coal distribution between burners	Emissions, slagging and heat rate increase due to inefficient combustion
Secondary Air (Windbox)	Air distribution through windbox	Emissions, slagging and heat rate increase due to inefficient combustion
Coal Pulverizers	Coal fineness	Lost generation due to unit derate
	Coal throughput	
	Pressure loss	Lost generation due to unit derate or increased energy usage by fans
NO_x Reduction — Overfire Air, Low NO_x burners	Effectiveness, influence on LOI and reducing atmospheres	Potential EPA fines, water wall tube replacement/repair, unburned carbon, heat rate increase due to higher exhaust temperatures
SCR Systems, SNCR Systems	Effectiveness, influence of gas distribution on NO_x reduction capability, additional pressure loss	Potential EPA fines, increased energy usage by fans
Fuel Blending/Substitution	Effectiveness	Slagging, heat rate increase, loss generation due to unit derate
Flue Gas Desulphurisation	Effectiveness, influence of gas distribution on SO_x reduction capability, additional pressure loss	Potential EPA fines, increased energy usage by fans
Various Tube Banks	Slagging, reduced heat transfer	Heat rate increase, lost generation due to derate and unit outage for cleaning
	Erosion (tube wastage)	Lost generation due to outage, tube replacement
Air Heater	Pluggage by flyash	Lost generation due to outage, water wash cost
Gas Recirc. System	Effect on steam temps, gas velocities through tube banks	Fan maintenance and power requirement, tube erosion and failure
Electrostatic Precipitator	Flow distribution, ash capture, ash dunes, gas conditioner distribution	Potential EPA particulate emissions fines, unit derates due to opacity
Baghouse	Temperature flow distribution, bag life	Premature bag replacement, fan requirements
Misc Ductwork	Pressure loss	Increased energy usage by fans, decreased gas turbine output
CEMS	Positioning for accurate readings	Loss emissions allowances, failed RTA tests

Table 1 - Economic Issues Associated with Selected Areas of Fossil-Fuelled Power Plants

192

Figure 1 - Cost of Unit Derate

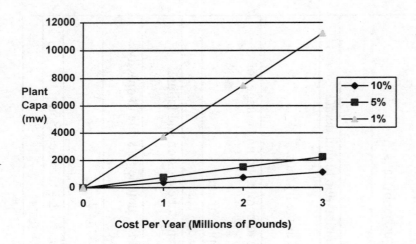

Figure 2 Cost of Net Plant Heat Rate Increase

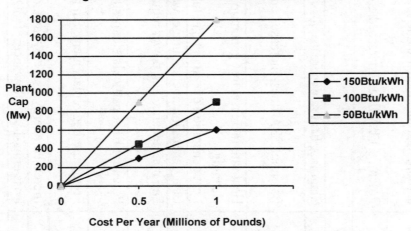

Figure 3 Cost of Increased Carbon in Ash

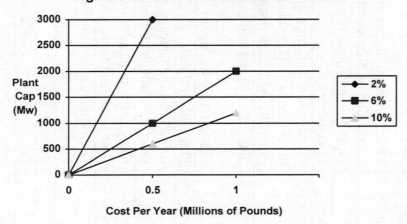

Figure 4 Cost of EPA NO$_x$ Emissions Fines

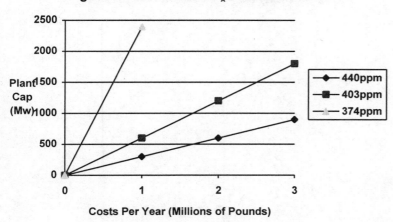

Figure 5 Capital Costs of NO$_x$ Control Systems

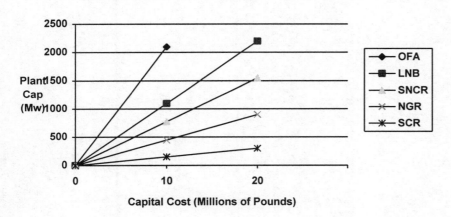